T0297989

Mathematical Modelling with Differential Equations

Mathematical Modelling with Differential Equations

Ronald E. Mickens
Clark Atlanta University, USA

CRC Press
Taylor & Francis Group
Boca Raton London New York

CRC Press is an imprint of the
Taylor & Francis Group, an **informa** business
A CHAPMAN & HALL BOOK

First edition published 2022
by CRC Press
6000 Broken Sound Parkway NW, Suite 300, Boca Raton, FL 33487-2742

and by CRC Press
4 Park Square, Milton Park, Abingdon, Oxon, OX14 4RN

CRC Press is an imprint of Taylor & Francis Group, LLC

ISBN: 978-1-032-01445-6 (hbk)
ISBN: 978-1-032-01530-9 (pbk)
ISBN: 978-1-003-17897-2 (ebk)

DOI: 10.1201/9781003178972

Typeset in Nimbus
by KnowledgeWorks Global Ltd.

Publisher's note: This book has been prepared from camera-ready copy provided by the authors.

Contents

Preface

This book is concerned with various strategies for modelling systems using, for the most part, differential equations. Some of these methodologies are elementary and quite direct to comprehend and apply while others are complex in nature and require thoughtful, deep contemplation. This volume is my version as to how the selected topics should be presented and modelled. Consequently, techniques will be introduced (sometimes) that are new, novel, unorthodox, and unconventional.

All of the topics have risen out of a three-semester course that I have taught at Clark Atlanta University during the past four decades (1983 – 2019). This series of courses, fulfilled in part the Physis Department's undergraduate and graduate programs for the BS, BS/MS, and MS degrees. The majority of the enrolled students came from the Physics and Mathematical Sciences Department. The essential prerequisites for these courses are the standard sequence of courses in beginning calculus, with additional courses in linear and matrix algebra, and differential equations.

In general, this book does not concern itself with mathematical theorem statements and proofs. We use mathematics mainly as a heuristic tool to provide insights into a given model. This means that our style of presentation and the possible interpretations that arise for a particular model are heavily based on physical and mathematical intuitions.

The individual chapters contain relatively few homework problems in comparison to standard texts on mathematical modelling. However, while some problems are easy, maybe even trivial to see/use/resolve, others are worthy of being topics for advanced degrees. We also make an attempt to provide references to the appropriate pedagogical and research literature.

The central concept holding the book together as a logical entity is the "principle of dynamic consistency." This idea is explained in detail in Chapter 0, section 0.2. It is of critical importance that Chapter 0 be thoroughly read and comprehended by users of this book, since it contains many useful topics and issues not generally found in standard books, and likewise, introduces several concepts and notions that are relevant to the productive use of this volume as a textbook and source for self-study. Note that each chapter can stand on its own merits; consequently, except for Chapter 0, the other chapters may be tackled in any desired order.

I hope you find this book challenging, but at the same time, interesting. Please communicate your thought to me regarding any aspect of this volume: methodology, topics, and presentation style.

Finally, I want to thank Callum Fraser, my Editor, who allowed me complete freedom to write this book in the unconventional style and format you are accessing.

Ronald E. Mickens
November 3, 2021
Atlanta, Georgia USA

CHAPTER 0

Preliminaries

0.1 INTRODUCTION

The main purpose of this chapter is to introduce and discuss the concept of a system, its mathematical modelling, and the value and usefulness of the principle of dynamic consistency. All of these items are illustrated by the consideration of several elementary systems and the resolution of certain questions concerning their properties and/or other features of interest.

Since all modelling is done within the context and constraints of science (physics), we devote Section 0.4 to explaining just what is science. This is followed, in Section 0.5, by a discussion on scaling of variables, using elementary dimensional analysis, and how this technique can often be used to both simplify the obtaining of an "answers," while also giving insight into the nature of the obtained results.

Section 0.6 examines the concept of dominant balance and the construction of valid approximations to the original modelling equations, which for most situations are differential equations.

The next two sections, 0.7 and 0.8, briefly introduce the look-up methodology using standard mathematical handbooks and Wikipedia listings. These techniques often allow us to rapidly determine what is currently known about a particular topic or issue and may point to additional sources of information.

Finally, the chapter concludes with additional comments on the modelling process.

0.2 MATHEMATICAL MODELLING

A system is a set of interrelated, interacting, or interdependent elements which together form a complex whole.

While this definition is vague, in general, when we see, construct or examine such an "entity," its "system nature" is almost always obvious to us. Also, note that a system might be conceptional, i.e., composed of non-physical objects or components, such as ideas or even mathematical expressions.

Definition 1: **Modelling** is the representation of one system by a second system.

Definition 2: **Mathematical modelling** is the representation of a system by a set of mathematical relations or equations.

DOI: 10.1201/9781003178972-0

It should be clear that for a given system, there is, in general, a large number of mathematical models that may be used to represent it. Thus, the mathematical modelling of a particular system is not unique.

To proceed, we introduce the concept of dynamic consistency.

Definition 3: Consider two systems S and $\bar{S}$, and let S have the property or feature P. If $\bar{S}$ also has property P, then $\bar{S}$ is said to be **dynamical consistent** with S with respect to P.

To illustrate dynamic consistency, we examine a physical system, S which has the following properties/features:

(i) It can be characterized by either particle numbers or a particle density.

(ii) The system satisfies a conservation law, such as constant total mass.

(iii) The system has several equilibrium states.

(iv) The system can undergo a number of phase-transitions.

For the system $\bar{S}$, we consider a mathematical model, $\bar{S}$, based on differential equations. For $\bar{S}$ to be dynamic consistent with respect to the four properties listed above for S, then we must have the following features for $\bar{S}$:

(a) The relevant solutions to $\bar{S}$ must be non-negative.

(b) The dependent variables in the differential equations must satisfy a mathematical version of the physical conservation law.

(c) The solutions of $\bar{S}$ must have fixed-points corresponding to the equilibrium states of the system S.

(d) The structure of the differential equations should have a form such that bifurcations occur at the phase-transition.

At this point, several additional comments need to be made. First, the system S will also have properties/features not listed above. Second, the mathematical model $\bar{S}$ may not include some (or any) of these additional properties/features. Thus, a hierarchy of mathematical models may be constructed, with higher level models including a larger number of these properties/features. The following might constitute one such hierarchy:

$$
\begin{aligned}
S &: (P_1, P_2, P_3, P_4, \ldots P_n, \ldots) \\
\bar{S}_1 &: (P_1, P_3, P_4) \\
\bar{S}_2 &: (P_1, P_2, P_3, P_4) \\
\bar{S}_3 &: (P_1, P_2, P_3, P_4, P_5) \\
\bar{S}_4 &: (P_1, P_2, P_3, P_4, P_6, P_8, P_9)
\end{aligned}
$$

Generally, it is expected that $\bar{S}$'s higher in the hierachy will provide better, more accurate representations or descriptions of S.

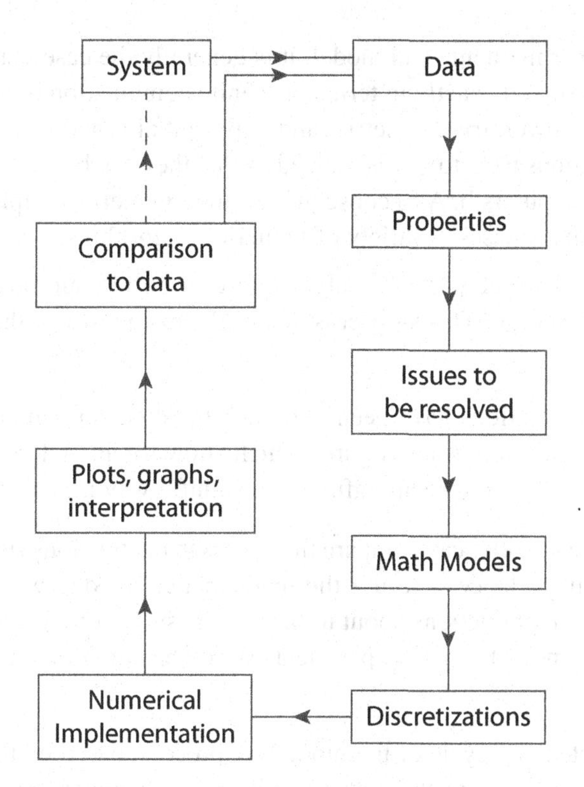

Figure 0.1: The full modelling process cycle.

As our final topic, we examine the overall modelling methodology. The full modelling process cycle is given in Figure 0.1.

In more detail, the modelling cycle processes as follows:

(1) First, the system is defined and all relevant independent and dependent variables, as well as the parameters, are identified.

(2) The system is then "probed" to generate a data set. This may be done by experimentation and/or by means of "intellectual analysis."

(3) From the data set, patterns are discovered and from these patterns, the major properties of the systems are determined.

(4) Given the results in (3), the important and critical issues that require resolutions are formulated.

(5) At this stage, generally there is sufficient knowledge of the system and its properties so that mathematical models may be formulated, constructed, and analyzed.

Note that the type of mathematical model to be used is not necessarily unique. For example, the mathematics used to model a particular system might be in terms of differential equations or difference equations or integral equations. The selection of a particular mathematical type is a function of the modeler's knowledge and experience.

(6) Given a particular mathematical model, it is generally the case that the relevant equations cannot be solved exactly in terms of a finite combination of the elementary standard functions: power, trigonometric, and exponential functions. This means that the modelling equations have to be discretized so that they can be studied by means of digital devices ("computers"). As a consequence, their numerical implementation must be done using, in many cases, a variety of numerical algorithms.

A central issue is which particular algorithm to use to obtain numerical solutions to the modelling equation. This subject is one of the main tasks of the field of numerical analysis.

(7) Once numerical solutions have been obtained, the modeler can analyze these results and construct plots, graphs, and figures which allow for an understanding of the original system. To do this in a meaningful and insightful way may be very difficult.

(8) At this stage, the modeler can compare the results in the previous step to the actual data. Also, of great importance is to use the mathematical model, with the corresponding numerics, to make predictions about unmeasured aspects of the system. These actions may suggest the need for new experiments or to change in the original mathematical model.

(9) Having completed one cycle of the modelling process, based on the comments in (8), there may be a need to repeat the cycle until a suitable understanding of the system is reached.

0.3 ELEMENTARY MODELLING EXAMPLES

To gain some understanding as to the power of the modelling process, we consider four "problems" that are resolved by modelling them with the use of elementary mathematics. Note that while these problems are elementary, they are not, for many, simple and straightforward to analyze.

0.3.1 Ball and Bat Prices

A ball and bat together cost $1.10. The bat costs $1.00 more than the ball. How much does the ball cost?

This question was asked by me to seven individuals and two possible answers were given. Five replied with the price of the ball being ten cents and, consequently, the bat is $1.00 or 100 cents. Two stated that the ball's price is five cents, while the bat's price is 105 cents. (Since we are using American currency, we have $1.00 equal to 100 cents.) Which of these answers is correct? May be, neither is right.

To resolve this issue, we convert this problem to one involving a mathematical representation of the items. First, in words, we have

(Price of bat) + (Price of ball) = 110 cents,

(Price of bat) − (Price of ball) = 100 cents.

Adding these two expressions gives

$$2\,(\text{Price of bat}) = 210 \text{ cents}$$

or

$$\text{Price of bat} = 105 \text{ cents}$$

From this, it follows that the price of the ball is five cents.

If we use mathematical symbols where

$$x = \text{Price of ball}, \;\; y = \text{Price of bat},$$

then our situation may be modelled using the two equations

$$\begin{aligned} y + x &= 110, \\ y - x &= 100. \end{aligned}$$

This system of two linear equations has the solution $x = 5$ and $y = 105$.

The elementary mathematical model gives an unambiguous, unique answer.

0.3.2 Planet-Rope Problem

Suppose you live on a planet of radius R. Place a rope around the planet on its equaters. Now increase the length of the rope by 10 meters, while keeping it in the plane of the equator. Question: How far above the surface of the planet is the lengthen rope?

Most individuals would guess that the rope would lie some small fraction of a meter above the surface of the planet given the fact that the radii of planets are huge compared to the 10 meter increase in the length of the rope. Again, this issue can be settled by the construction of a mathematical model of the problem.

Make the following definitions:

R = radius of planet,

h = height of the lengthen rope above the surface of the planet.

From elementary geometry, we have for a circle

$$C = 2\pi r,$$

i.e., the circumference of the circle is equal to 2π ($\pi \approx 3.1416$) times it radius.

For our situation, we have (see Figure 0.3.1)

$$\begin{aligned} C(\text{planet}) &= 2\pi R, \\ C(\text{rope}) &= 2\pi(R + h). \end{aligned}$$

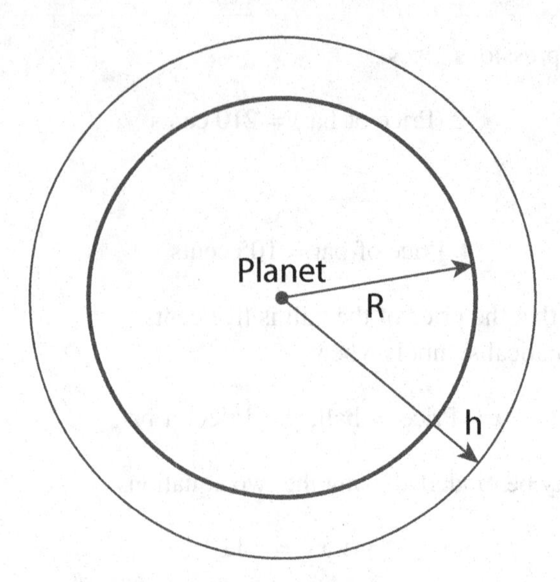

Figure 0.3.1: Planet surrounded by a rope at a distance h from its surface.

However, note that

$$C(\text{rope}) - C(\text{planet}) = 10 \text{ meters},$$

Since the lengthen rope is 10 meters longer than the circumference of the planet. Therefore,

$$2\pi(R + h) - 2\pi R = 10,$$

or

$$2\pi h = 10,$$

and solving for h gives

$$h = \frac{10}{2\pi} \approx 1.59 \text{ meters}.$$

This is a very surprising result: first, because the height above the surface of the planet does not depend on the radius of the planet; second, its value is approximately 1.6 meters, a magnitude that we might have expected to be much, much smaller.

This example clearly illustrates the power of mathematical modelling.

0.3.3 The Traveler and the Mountain Climb

A traveler to the country of Xezena wishes to climb its only mountain. The country is flat, except for the mountain which has a peak of height h (see Figure 0.3.2).

At sunrise of the first day, the traveler starts at A, at the foot of the mountain, and climbs to its peak, B. This is done in a monotonic manner, i.e., the climbing and rest stops are no where a decreasing distances of the height from the plain. After sightseeing and rest for two days, the traveler decides to return to the foot of the mountain, A, along a path from

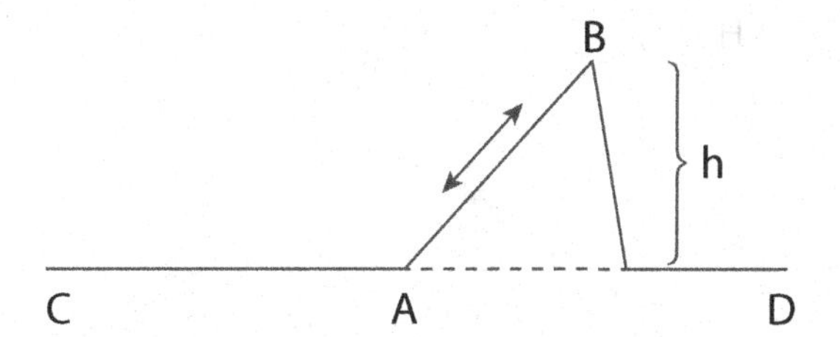

Figure 0.3.2: Mountain rising above the plain CD. The traveler ascends from A to B and later descends from B to A.

B to A. Again, the path is taken to be a monotonic decrease in terms of the height above the plain. The traveler begins the descend at sunrise.

Define $H(u,t)$ and $H(d,t)$ to be

$$\begin{aligned} H(u,t) &= \text{distance above the plain on the ascend to the peak,} \\ H(d,t) &= \text{distance above the plain on the descend to the plain,} \\ t &= \text{elapsed time from the beginning of either the ascend or descend.} \end{aligned}$$

Figure 0.3.3 are plots of $H(u,t)$ and $H(d,t)$ as a function of t.

The question of interest is whether there is a single point along the path for which it is reached at exactly the same time for both the climb up the mountain and the climb down the mountain?

This issue can be easily resolved by combining the two graphs in Figure 0.3.3 and plotting them together. This is done and shown in Figure 0.3.4.

First, observe that t_{BA}, the time to climb down the mountain, may not be the same as the time to climb to its peak, t_{AB}. But, under the assumption of monoticity there is a unique time, t^*, where the two curves intersect. Thus, the above question is answered in the affirmative, i.e., there is a single point along the path for which the traveler would pass at exactly the same time for both the ascend and descend of the mountain.

An interesting feature of this problem is that an answer can be produced by using the appropriate drawing. No numerics or formulas need to be introduced.

0.3.4 What is the Sum: $\frac{1}{2} + \frac{1}{4} + \frac{1}{8} + \cdots$?

The following physical experiment can actually be done (up to a point).

(i) Take a sheet of paper and cut it in half.

(ii) Place one half on a table and cut the other half into two equal parts. Now put the half of a half piece of paper with the piece of paper on the table.

(iii) Repeat this procedure, always placing the newly cut half on the table with the previous cuts.

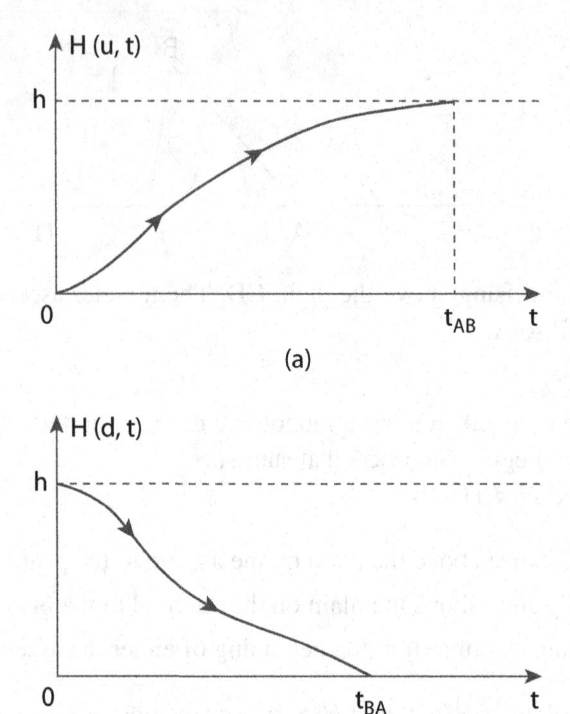

Figure 0.3.3: Plots of (a) distance above the plain on the way to the mountain peak versus time elapsed since the beginning of the ascend; (b) distance above the plain for a starting at the mountain top and descending to the plain.

If the original sheet of paper has an area of one unit and if the collective area of the paper cuts on the table after N cuts is denoted S_N, it follows that

$$
\begin{aligned}
S_1 &= \frac{1}{2} = 0.500000 \\
S_2 &= \frac{1}{2} + \frac{1}{4} = 0.750000 \\
S_3 &= \frac{1}{2} + \frac{1}{4} + \frac{1}{8} = 0.875000 \\
S_4 &= \frac{1}{2} + \frac{1}{4} + \frac{1}{8} + \frac{1}{16} = 0.937500 \\
S_5 &= \frac{1}{2} + \frac{1}{4} + \frac{1}{8} + \frac{1}{16} + \frac{1}{32} = 0.968750 \\
S_6 &= \frac{1}{2} + \frac{1}{4} + \frac{1}{8} + \frac{1}{16} + \frac{1}{32} + \frac{1}{64} = 0.984375 \\
&\vdots \qquad\qquad \vdots \\
S_N &= \frac{1}{2} + \frac{1}{4} + \cdots + \frac{1}{2^N} \\
&\vdots \qquad\qquad \vdots
\end{aligned}
$$

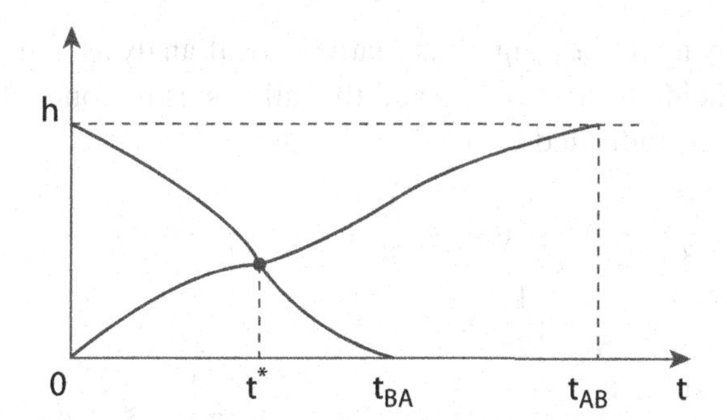

Figure 0.3.4: Plots of $H(d, t)$ and $H(u, t)$ on the same graph. The time variable t stands for the time from the initial event on each graph to its completion.

So what is S_∞, i.e., the area of the combined "one-half" cut pieces after an infinite number of cuts?

Since the original whole sheet had an area of one and assuming that we can physically continue to do the "half cutting" indifferently and placing one of the newly cut pieces on the pile on the table, we reach the **experimentally based conclusion** that

$$S_\infty = \lim_{N\to\infty} S_N = 1.$$

Therefore, we conclude that

$$\frac{1}{2} + \frac{1}{4} + \frac{1}{8} + \frac{1}{16} + \cdots = 1,$$

where the $(\cdots)$ indicates all the other terms in sum. Again, this constitutes a physical proof of the value of the sum. Thus, a mathematical issue has been modelled by a physical process.

Another way of proceeding is to replace $\frac{1}{2}$ by the symbol X and obtain

$$S_\infty = S(x) = x + x^2 + x^3 + \cdots + x^k + \cdots \tag{0.3.1}$$

This expression contains a factor x in each term and can be rewritten as

$$\begin{aligned} S(x) &= x(1 + x + x^2 + \cdots + x^{k-1} + x^k + \cdots) \\ &= x[1 + S(x)]. \end{aligned}$$

Solving for $S(x)$ gives

$$S(x) = \frac{x}{1-x}. \tag{0.3.2}$$

If x is replaced by $x = \frac{1}{2}$, then the result $S(\frac{1}{2}) = 1$ is obtained, the value obtained from the "physical experiment."

Now $S(x)$ is defined for all $x \neq 1$ and examination of the expression in Eq. (0.3.1) shows that for $x > 0$, S_∞ is positive, i.e., it is the sum of positive terms. However, $S(x)$, as presented in Eq. (0.3.2), is negative for $x > 1$ and $x < 0$. Also, note that $S(1)$ is not defined. Therefore, we limit our use of $S(x)$ to the interval $0 \leq x < 1$.

So, without the benefits of a rigorous mathematical analysis, we assume that the formula of Eq. (0.3.2) holds for $0 \leq x < 0$ and this allows us to conclude that the following sums exist and have the indicated values:

$$\frac{1}{3} + \frac{1}{9} + \frac{1}{27} + \cdots + \frac{1}{3^k} + \cdots = \frac{1}{2},$$
$$\frac{1}{5} + \frac{1}{25} + \frac{1}{125} + \cdots + \frac{1}{5^k} + \cdots = \frac{1}{4},$$
$$\frac{1}{\pi} + \frac{1}{\pi^2} + \frac{1}{\pi^3} + \cdots + \frac{1}{\pi^k} + \cdots = \frac{1}{\pi - 1} \approx 0.466942.$$

0.4 WHAT IS SCIENCE?

Perhaps the beginning of science started when we as humans realized and understood that ignorance existed, i.e., there are things that we do not know, but we can do something about this situation. Some particular examples of phenomena for which early humans sought explanations were:

(a) The daily "rising" and "setting" of the sun;

(b) What is heat? And why does water get hotter when heat is applied to it?

(c) Why does $3^2 + 4^2 = 5^2$? And what are its implications?

The common feature to all these situations is the existence of certain patterns and this may be expressed as follows: If a set of conditions A hold, then conditions B must also hold. This "physical" collection of observations, along with "mental" analysis leads to the following definition of science:

> Science **is the systematic observation, creation, analysis, and modelling of patterns which exist in the physical world.**

In a similar manner, mathematics may be defined as

> Mathematics **is the study, creation, and analysis of patterns in the abstract universe of human thought and mental perception.**

Accepting these two definitions allows for the resolution of the problem as to why mathematics can be considered the "language of science." Namely, both subjects deal with patterns and mathematics has the flexibility and the possibility for generalization to allow it to be directly applicable to the task of interpreting, understanding, and predicting the structures of possible patterns which may occur in the physical universe. Thus, mathematics can function as a language for the construction of physical theories which, in turn, may be used to provide scientific explanation of the natural universe.

Another significant aspect of science is that it provides us with public knowledge, i.e., knowledge that is available for anyone to examine, test, criticize, and generalize. The public knowledge aspect of science insures that scientific facts have certain objective nature, i.e., if "I" perform an "experiment" and obtain certain results, then any other (competent) individual can also carry out this same experiment. Experience shows that in the absence of "major" errors by either party, the "same" results will be obtained.

Another important consequence of the public knowledge aspect of science is the lack of need for supreme authority figures, whose knowledge is derived in part from sacred of inspired thoughts or texts. It should be noted that much of general knowledge is not directly related to or based on scientific knowledge. However, it is also not "non-sense" either. Not all personally valuable and worthwile knowledge is scientific. For example, consider poetry. It is clearly not scientific knowledge: different individuals, if requested to write a poem on love, will almost certainly produce entirely different poems, based on their widely varying life experiences, writing abilities, creativity, etc. But who would wish to give up the construction and readings of these works?

Science is also a special type of knowledge, having a very strict methodology. However, there is no general scientific method. If there exists a scientific procedure, its statement would be: try everything, but don't be stupid.

Another critical aspect of science is that "Science is not democratic or fair. The empirical evidence and logical critical analysis rules" (Flammer, 2015).

Thus, the general influence of statements such as "my opinion is ... " or the pronouncements of any type of authorities have no long term impacts on scientific knowledge and progress.

Finally, science is a progressive field of inquiry and, in general, does not place a high value on its prior history of achievements. For example, the Greek understanding of atoms was eventually replaced by Newtonian atoms, which in turn were replaced by Bohr atoms, which are presently interpreted as quantum atoms. But, to study and fully understand quantum atoms, there is no necessary need or requirement to study the history of atomic theory, starting with the Greeks. One may begin with the latest results from quantum theory and its calculational methodology (Simonyi, 2012). However, to gain a sense of the progression and progress of the atomic concept, then this history is important to know.

A measure of the validity of subjects or areas which want to be thought of as "being scientific" is the extent to which they discuss issues important to distance, earlier periods, in comparison to current problems. This situation demonstrates that they are not interested in making a scientific effort, but merely using their arguments to "verify" an **a priori** set of fixed beliefs.

0.4.1 Definitions, Comments, and Statements on Science and Related Issues

Abstraction: This is the process of removing characteristics from something that we do not need in order to reduce it to a set of essential features.

Antropic Principal: Scientific theories must be constructed such that the structure of the physical universe is compatible with the existence of humans as observers.

Applied and Mathematics: Pure mathematics are those areas of mathematics that have not found serious "physical" applications outside of mathematics itself. Applied mathematics is mathematics that has found such applications.

Apophenia: This is the human tendency to perceive meaningful patterns within random data.

Approximation: To approximate something is to represent it with something else that is good enough for your purpose.

Bad Result: A result which would have been quickly forgotten had it been remembered in the first place.

Broredom: This is the uncomfortable experience of wanting to engage in a satisfying activity, but being unable to do so. Boredom generally arises when there is some constraint, from monotony, the lack of purpose, and the non-existence of a meaningful challenge.

Childhood: Childhood is similar to being drunk ... everyone remembers what you did except you.

Common Sense: This is the set and interpretation of everyday experiences that are encountered by the normal processes of living.

Conjecture: A conjecture is thoughtful guess that is based on all the known relevant information. In general, this is a concisely formulated statement that is likely to be true, but for which no conclusive proof has yet been found.

Convergent Series: In actual practice, mathematicians assume that a given series diverges, until proven otherwise. However, a physical scientist assumes a series converges until proven otherwise and then reinterprets the series to make physical sense.

Euler-Mickens Day: I have been told, many times by my mother and grandmother that I was born on February 7th, at 6:28 PM, in Petersburg, Virginia. Using the 24-hour notation, this has the representation

$$\text{month/day/hour/minute} = 2\text{-}7\text{-}18\text{-}28$$

Now, it is of interest to note that the Euler number is

$$e = 2.71828\ldots$$

Just like the Pi Day celebration on March 14, I am suggesting that we also celebrate the Euler-Mickens Day during the late afternoon each February 7.

Experiment and Measurement: "An experiment is a question which science poses to nature, and a measurement is the recording of nature's answer. " (Max Planck)

Geniuses: Mark Kac explains that there are two types of geniuses. The ordinary geniuses "is a fellow that you and I would be just as good as, if we were only many times better. There is no mystery as to how there minds work. Once we understand what he has done, we feel certain that we, too, could have done it." The second type of geniuses are the "magicians ... Even after we understand what they have done, the process by which they have done it is completely dark."

IQ Fairy: I am aware that several of my colleagues never had a visit from the "IQ Fairy."

Idea Zombies: These are bad ideas that will not stay in their intellectual graves.

Ignorance and Stupidity: Ignorance is the lack of knowledge and understanding. It can be cured by education. Stupidity is the inability to be ignorant.

Invisible College: The invisible college concept provides a possible structure for the organization of scientists in a particular discipline. It is a set of interacting scientists who share similar research interests concerning a subject specialty, who produce publications relevant to this subject and who communicate both formally and informally with one another to work toward important goals in the subject, even though they may belong to geographically distant research affiiliates.

Life and Death: Life is one story after another. Death is the end of the story.

New Scientific Truth: "A new scientific truth does not triumph by convincing its opponents and making them see the light, but rather because it opponents eventually die, and a new generation grows up that is familiar with it." (Max Planck)

Order of Magnitude: A way to define the "order of magnitude" of a real (positive) number N is to represent it in the form

$$N = a \cdot 10^b,$$

where

$$\frac{1}{\sqrt{10}} \leq a < \sqrt{10}, \quad b = \text{integer}.$$

The say that N is of order of magnitude 10^b.

Oxygen: Most people consider oxygen to be of benefit to humans. But it may separation be that oxygen is slowly killing us and it takes 75–100 years to do so.

Philosophy: The study of important issues that cannot be resolved through scientific methodology alone.

Pseudoscience vs Science: The demarcation line between science and pseudoscience is testability.

Research: is the act of trying to find the answers to interesting questions that in principle can be answered.

Science: Science is the methodology that we use to uncover the "rules" governing the physical universe. The domain of science is not the set of things that we already know, that is knowledge. The domain of science is everything we do not know.

Scientist: A scientist is someone who systematically gathers and uses (experimental and analytical) research techniques to formulate hypotheses and test them, and to gain and share their understandings and knowledge. Any, when possible or appropriate generalize these results.

Success: A person's success in life is determined by having a high minimum, not a high maximum.

Thought Experiment: Thought experiments use the mind to create rational situations obeying the known laws of physics and the rules of logic.

Toy Models in Science: A toy model is a deliberately constructed simplistic model with inessential details not inclused so that it may be used to analyze or explain some mechanism or feature of a particular system concisely. A toy model may be used as the starting point for the construction of a hierarchy of more complex models.

0.5 SCALING OF VARIABLES

0.5.1 Mathematical and Physical Equations

The simple, damped harmonic oscillator can be modelled by the following linear, second-order differential equation

$$m\frac{d^2x}{dt^2} + K_i\frac{dx}{dt} + K_2x = 0. \tag{0.5.1}$$

This equation is a consequence of the application of Newton's force law to a mass, m, attached to a linear spring, where the force exerted by the spring is $(-K_2x)$, and a linear damping also acts, F(damping) $= -K_i dx/dt$. The variable x represents the displacement of the mass from its equilibrium position at $x = 0$. This system has three parameters, namely, (m, k_1, k_2), where k_1 and k_2, are respectively, the damping and spring constants.

In the context of classical point mechanics, there are three fundamental physical units given below.

Within the so-called MKS system, these units have the indicated values and are generally abbreviated as m-meter, kg-kilogram, and s-second. Every physical variable has a definite set of physical units or dimensions and some of the common ones are listed below.

length	L	meter
mass	M	kilogram
time	T	second

Quantity	Dimension
area	L^2
volume	L^3
velocity	L/T
acceleration	L/T^2
momentum	ML/T
force	ML/T^2
energy	ML^2/T^2
pressure	M/LT^2
power	ML^2/T^3
mass density	M/L^3

Returning to Equation (0.5.1), every term has the dimensions of force. This is obvious for the first term. The dimensionality of the parameters k_1 and k_2 may be determined using the fact that

$$\left[k_i\frac{dx}{dt}\right] = [k_1]\left[\frac{dx}{dt}\right] = \frac{ML}{T^2}, \tag{0.5.2}$$

$$[k_2x] = [k_2][x] = \frac{ML}{T^2}, \tag{0.5.3}$$

where $[\cdots]$ represents the dimensionality of the expression in the squae brackets.

Therefore, it follows that

$$[k_1] = \frac{M}{T}, [k_2] = \frac{M}{T^2}. \tag{0.5.4}$$

From k_1 and k_2, two "time scales" can be constructed, i.e.,

$$T_1 = \frac{m}{k_1}, T_2 = \left(\frac{m}{k_2}\right)^{\frac{1}{2}}. \tag{0.5.5}$$

Note that T_1 is the characteristic damping time, while T_2 is the characteristic time for the free oscillations.

For the initial conditions

$$x(0) = A, dx(0)/dt = 0, \tag{0.5.6}$$

form the following two dimensionless variables

$$\bar{x}(t) = \frac{x(t)}{A}, \bar{t} = \frac{t}{T_2}. \tag{0.5.7}$$

Substituting these expressions into Equation (0.5.1) and simplifying gives

$$\begin{cases} \dfrac{d^2\bar{x}}{d\bar{t}^2} + 2\epsilon\dfrac{d\bar{x}}{d\bar{t}} + \bar{x} = 0, \\ \bar{x}(0) = 1, d\bar{x}(0)/d\bar{t} = 0, \end{cases} \tag{0.5.8}$$

where

$$2\epsilon = \frac{T_2}{T_1}. \tag{0.5.9}$$

In other words, ϵ is the ratio between the time associated with the free oscillations, when $k_1 = 0$, and the time associated with the damping.

Note that Eq. (0.5.8) has the following important features:

(i) All the variables and parameters in this differential equations are dimensionless and therefore can be represented by pure numbers.

(ii) This dimensionless form of Equation (0.5.1) has only one parameter in contrast to the original differential eqution which contains three parameters, (m, k_1, k_2).

(iii) Also, the dimensionless equation has only one set of values for the initial conditions, $\bar{x}(0) = 1$ and $d\bar{x}(0)/d\bar{t} = 0$, as compared with Eq. (0.5.1) which has an infinite possible set of initial values, $x(0) = A$ and $dx(0)/dt = 0$. This simplication of possible initial values is of great advantage for the purposes of numerical integration.

Equation (0.5.1) is called a **physical equation**, while the expression given in Equation (0.5.8) is called a **mathematical equation**.

Definition 1: A physical differential equation has, in general, all of its variables and parameters expressed in terms of a given set of physical units such as mass, length, and time.

Definition 2: **Mathematical equations** have all of their variables and parameters dimensionless, i.e., pure real numbers.

Mathematical equations may be considered models of physical equations for which all variables and parameters have been made dimensionless. These new equations generally are easier to investigate since they contain a smaller number of parameters and simpler initial conditions.

0.5.2 Characteristic Scales and Dimensionless Variables

One method for constructing dimensionless equations is to using the following technique:

(a) Let (u, x, t) be, respectively, the dependent variables, and space and time independent variables.

(b) Define new dimensionless variables by means of the relations

$$u = u^*\bar{u}, x = L\bar{x}, t = T\bar{t}, \tag{0.5.10}$$

where $(\bar{u}, \bar{x}, \bar{t})$ are the new dimensionless variables and (u^*, L, T) are the, at this stage, unknown scaling factors.

(c) If the set of differential equations take the form

$$\frac{du}{dt} = f(x,t,u,p), p = \text{parameters}, \tag{0.5.11}$$

then upon making the substitutions of Equation (0.5.10) into it, we obtain

$$\frac{d\bar{u}}{d\bar{t}} = F(\bar{x},\bar{t},\bar{u},p), \tag{0.5.12}$$

where the function F is fully determined from a knowledge of f.

(d) Since the left-side of this equation is dimensionless, then all of the separate terms in F must be dimensionless.

(e) By selectively equating the coefficients of these terms to zero, the scales, (u^*, L, T) may be determined.

(d) Note that the particular value choosen for the coefficients is arbitrary. A choice is often made to select the value of one.

The next section illustrates this technique by applying it to three examples.

0.5.3 Examples of Scaling

0.5.3.1 Decay equation

This differential equation models the decay of an atomic nucleus as in radioactivity. It takes the form

$$\frac{dc}{dt} = -\lambda c, c(0) = c_0 > 0, \tag{0.5.13}$$

where $c(t)$ has the physical units of atoms per unit volume. We have

$$c = c^*\bar{c}, t = T\bar{t}, \tag{0.5.14}$$

which upon substituting in the differential equation gives

$$\left(\frac{C^*}{T}\right)\frac{d\bar{c}}{d\bar{t}} = -(\lambda c^*)\bar{c}, c^*\bar{c}(0) = c_0, \tag{0.5.15}$$

or

$$\frac{d\bar{c}}{d\bar{t}} = -(\lambda T)\bar{c}, \bar{c}(0) = \frac{c_0}{c^*}. \tag{0.5.16}$$

If the following selections are made

$$T = \frac{1}{\lambda}, c^* = c_0, \tag{0.5.17}$$

then the mathematical equation for the decay equation is

$$\frac{d\bar{c}}{d\bar{t}} = -\bar{c}, \bar{c}(0) = 1, \tag{0.5.18}$$

This differential equation and its initial condition has no free parameters.

0.5.3.2 Duffings equation

This equation models a nonlinear mechanical oscillator and its mathematical structure takes the form of a nonlinear, second-order, ordinary differential equation

$$m\frac{d^2x}{dt^2} + Kx + K_1x^3 = 0. \tag{0.5.19}$$

We select the initial conditions to be

$$X(0) = A, \frac{dx(0)}{dt} = 0. \tag{0.5.20}$$

This system has three parameters, (m, k, k_1), and one nontrivial initial condition, $x(0) = A$. These items have the physical units

$$\begin{cases} [m] = M, [k] = \frac{M}{T^2}, [k_1] = \frac{M}{L^2T^2}, \\ [x] = L, [A] = L. \end{cases} \tag{0.5.21}$$

If we choose our scaled variables to be

$$x = x^*\bar{x}, t = T\bar{t}, \tag{0.5.22}$$

it seems reasonable to take $x^* = A$. Substituting these results into Eq. (0.5.19) gives

$$m\left(\frac{A}{T^2}\right)\frac{d^2\bar{x}}{d\bar{t}^2} + (KA)\bar{X} + (k_1A^3)\bar{x}^3 = 0. \tag{0.5.23}$$

and upon multiplying by T^2/mA gives

$$\frac{d^2\bar{x}}{d\bar{t}^2} + \left(\frac{kT^2}{m}\right)\bar{x} + \left(\frac{k_1A^2T^2}{m}\right)\bar{x}^3 = 0. \tag{0.5.24}$$

Now set the coefficient of $\bar{x}$ equal to one and solve for the time scale, T; doing this gives

$$T = \sqrt{\frac{m}{k}}, \tag{0.5.25}$$

which is the characteristic time of the free oscillations for the linear part of this system. If this value for T is substituted into the coefficient of the $\bar{x}^3$ term then we obtain the result

$$\begin{aligned} \left(\frac{k_1A^2}{m}\right)T^2 &= \left(\frac{k_1A^2}{m}\right)\left(\frac{m}{k}\right) \\ &= \frac{A^2}{L_1^2}, \end{aligned} \tag{0.5.26}$$

where L_1 is an additional scale for $x(t)$, equal to

$$L_1 = \sqrt{\frac{k}{k_1}}. \tag{0.5.27}$$

Note that L_1 depends only on the parameters k and k_1, and this means that it is a length scale inherent to the system itself. If the dimensionless parameter, ϵ, is defined to be

$$\epsilon = \left(\frac{A}{L_1}\right)^2, \tag{0.5.28}$$

then the mathematical equation corresponding to the Duffing Equation (0.5.19) is

$$\frac{d^2\bar{x}}{d\bar{t}^2} + \bar{x} + \epsilon\bar{x}^3 = 0, \bar{x}(0) = 1. \tag{0.5.29}$$

Again, observe that while the physical equation has four parameters, (m, k, k_1, A), the mathematical, dimensionless equation has only one parameter, ϵ.

0.5.3.3 Fisher Equation

This nonlinear partial differential equation in $u = u(x, t)$, is

$$\frac{\partial u}{\partial t} = D\frac{\partial^2 u}{\partial x^2} + \lambda_1 u - \lambda_2 u^2, \tag{0.5.30}$$

where the parameters, $(D, \lambda_1, \lambda_2)$, are all non-negative. If the physical unit of u is particle density, then

$$[u] = \#, [D] = \frac{L^2}{T}, [\lambda_1] = \frac{1}{T}, [\lambda_2] = \frac{1}{\#T}. \tag{0.5.31}$$

The scales are (u^*, T_1, L_1) and we have

$$u = u^*\bar{u}, t = T_1\bar{t}, X = L_1, \bar{x}, \tag{0.5.32}$$

where, after a little calculation, they are determined to be

$$u^* = \frac{\lambda_1}{\lambda_2}, T_1 = \frac{1}{\lambda_1}, L_1 = \sqrt{\frac{D}{\lambda_1}}. \tag{0.5.33}$$

Therefore, the mathematical equation corresponding to the physical Fisher equation is, in the dimension less variables,

$$\frac{\partial \bar{u}}{\partial \bar{t}} = \frac{\partial^2 \bar{u}}{\partial \bar{x}^2} + \bar{u}(1 - \bar{u}). \tag{0.5.34}$$

Since there exist characteristic length and time scales, a characteristic velocity scale can be calculated, it is

$$c^* = \frac{L_1}{T_1} = (\lambda_1 D)^{\frac{1}{2}}, \tag{0.5.35}$$

and permits an estimate of how fast phenomena modelled by the Fisher equation propagate. A detailed examination of this equation shows that the minimum speed of propagation for non-negative solutions is

$$c = 2(\lambda, D)^{\frac{1}{2}}. \tag{0.5.36}$$

0.6 DOMINANT BALANCE AND APPROXIMATIONS

0.6.1 Dominant Balance

Once a mathematical model has been constructed for a particular system, it is rare that the equations can be solved exactly in terms of a finite combination of elementary functions. Thus, approximations to the actual solutions must be determined. In general, there are many methods available for carrying out this task. One importance technique is the method of **dominant balance**.

This procedure can be summarized as follows:

(1) Assume that the original equation consists of several terms; for example,

$$A + B + C + D + E = 0. \tag{0.6.1}$$

(2) Assume that two of the terms are "much larger" than the other terms. (In general, what terms are "much larger" and how they are characterized as such is dependent on the equation under study.) Take these two terms to be B and D.

(3) Now approximate the full original equation by the reduced equation

$$B + D = 0, \tag{0.6.2}$$

and solve it.

(4) Finally, check that the reduced equation is **consistent** with the original equation by substituting its solution into the original equation and determining that the neglected terms (A,C,E) are much smaller than the two kept terms. If this is not the case, the approximate solution is inconsistent and another method must be found, including the use of a different set of terms in the original equations.

Comments

It should be understood that there are no fully agreed on set of rules that determine the method of dominant balance. Also, for a given equation there may be several applicable dominant balance procedures. In actual practice this means that the modeller must have deep insights into the nature of the equations and their possible solution behaviors.

Another critical point to be noted is that the method of dominant balance is always associated with calculating approximate solutions for some particular asymptotic limit. Examples include independent variable(s) small,

- independent variable(s) small,
- independent variable(s) large,
- independent variable(s) having finite values.

0.6.2 Approximations of Functions

Often a function is provided and we would like to determine its behavior when one of its variables is small or large. For example, consider the function

$$f(x) = \frac{1}{1-x}. \tag{0.6.3}$$

It is defined for all values of x except for $x = +1$. Thus, for small x, we have

$$\begin{cases} f(x) = 1 + x + x^2 + \cdots + x^N + \cdots, \\ |x| < 1. \end{cases} \tag{0.6.4}$$

Now, what is its form for large x^2. We can determine this as follows:

$$\begin{aligned} f(x) &= (-1)\left[\frac{1}{x-1}\right], |x| > 1, \\ &= (-)\left[\frac{1}{x(1-\frac{1}{x})}\right] \\ &= (-)\left(\frac{1}{x}\right)\left[1 + \frac{1}{x} + \frac{1}{x^2} + \cdots + \frac{1}{x^N} + \cdots\right]. \end{aligned} \tag{0.6.5}$$

Note that these two expansions contain within them information to determine what is small x and what is large x. Basically, this is when the expansions make mathematical sense. Thus, $f(x)$, in Equation (0.6.4), is convergent when $|x| < 1$, and for x to be "small," we require $|x| << 1$, where the symbol "<<" means much smaller than. Likewise, since

$$\left|\frac{1}{x}\right| < 1, |x| > 1, \tag{0.6.6}$$

then large x is the requirement

$$|x| >> 1, \tag{0.6.7}$$

where the symbol ">>" means much greater than.

In practice, only a few terms in these expansions are required and we write them as expressions as

$$\text{x-small:} f(x) = \frac{1}{1-x} = 1 + x + O(x^2), \tag{0.6.8}$$

$$\text{x-large:} f(x) = \frac{1}{1-x} = \left(\frac{-1}{x}\right)\left[1 + \frac{1}{x} + O\left(\frac{1}{x^2}\right)\right]. \tag{0.6.9}$$

The notation O means the following:

The symbol O

Let $f(z)$ and $g(z)$ be functions of z. Let there exist a positive number A, independent of z, and a number $z_0 > 0$, such that

$$|f(z)| \leq A|g(z)|, \text{ for all } |z| \leq z_0, \tag{0.6.10}$$

then

$$f(z) = O[g(z)]. \tag{0.6.11}$$

It should be clear that asymptotic relations do not uniquely define a function, i.e., there are many functions that have exactly the same asymptotic representation. For example, for "small-x," we have

$$\left\{\begin{array}{l} \frac{1}{1-x} = 1 + x + O(x^2), \\ e^x = 1 + x + O(x^2), \\ \sin x = x + O(x^3), \\ \cos x = 1 + O(x^2). \end{array}\right\} \tag{0.6.12}$$

The next section gives examples of the use of dominant balance and the approximation of functions.

0.6.3 Examples

0.6.3.1 A damped harmonic oscillator

Consider the following second-order, linear differential equation

$$\ddot{x} + t\dot{x} + x = 0, \tag{0.6.13}$$

where the dot notation for time derivatives is used, i.e., $\dot{x} = dx/dt$ and $\ddot{x} = d^2x/dt^2$. This equation models a damped harmonic oscillator with the damping coefficient proportional to the time, t.

We wish to obtain information about the behavior of the solution for large times.

The nature of this physical system clearly indicates that the long time solution should decrease as a function of time.

To proceed, assume that this solution, i.e., the solution for large t, is a power law expression and takes the form

$$x(t) \sim At^{\alpha}, \tag{0.6.14}$$

where A and α are constants. Since our differential equation is a linear equation, the constant A can be arbitrary and we set it to the value one.

Thus, only α needs to be calculated.

Let us assume that $x(t)$, as given above in Equation (0.6.14), may be differentiated twice; therefore

$$x(t) \sim t^{\alpha}, \quad \dot{x}(t) \sim \alpha t^{\alpha-1}, \quad \ddot{x}(t) \sim \alpha(\alpha-1)t^{\alpha-2}. \tag{0.6.15}$$

Substitution of these results into Equation (0.6.13) gives the asymptotic relation

$$\alpha(\alpha-1)t^{\alpha-2} + \alpha t^{\alpha} + t^{\alpha} \sim 0,$$

or

$$\alpha(\alpha-1)t^{\alpha-2} + (\alpha+1)t^{\alpha} \sim 0. \tag{0.6.16}$$

Note that if $\alpha = -1$, then the second term is zero and the first term decreases as $1/t^3$, thus we have

$$t^{\alpha-2} = O(t^{-3}), t^{\alpha} = O(t^{-1}), \tag{0.6.17}$$

and the second and third terms dominant the first term in Eq. (0.6.13). If follows that the largest behavior of the solution is

$$x(t) \sim A/t. \tag{0.6.18}$$

Or is it? What has been shown is that the large t property given in Equation (0.6.19) holds if we assume the power law form expressed in Equation (0.6.14). What happens if we take a more general structure.

Assume that

$$x(t) \sim e^{-Bt^{\alpha}} \tag{0.6.19}$$

where α and B are positive constants. This form is a type of generalized exponential solution which decays to zero as $t \to \infty$.

Substitution into the different equation gives

$$-A\alpha(\alpha-1)t^{\alpha-2} + A^2\alpha^2t^{2(\alpha-1)} \tag{0.6.20}$$

$$-A\alpha t^{\alpha} + 1 \sim 0. \tag{0.6.21}$$

Matching the first and fourth terms, and the second and third terms gives, respectively,

$$\begin{cases} -A\alpha(\alpha-1) + 1 = 0, \alpha - 2 = 0, \\ A^2\alpha^2 - A\alpha = 0, 2(\alpha-1) = \alpha. \end{cases} \tag{0.6.22}$$

The common solutions are

$$\alpha = 2, A = \frac{1}{2}. \tag{0.6.23}$$

But, an examination of what we just did shows than an exact solution has been found, namely,

$$x(t) = Ae^{-\frac{t^2}{2}}, \tag{0.6.24}$$

where A is an arbitrary constant. Since the original differential equation is linear and of second-order, a second solution exists, but cannot be found using the current method.

What these calculations show is that one has to be very careful when applying dominant balance. If an approximation to the solution is obtained, it may have the correct qualitative properties, while not being an actual mathematical approximation to the exact solution.

0.6.3.2 Extension of a function

An elementary expression for the calculation of numerical solutions to the decay equation,

$$\frac{dx}{dt} = -\lambda x, x(0) = x_0 > 0, \tag{0.6.25}$$

is the following forward-Euler scheme

$$\frac{x_{k+1} - x_k}{h} = -\lambda x_k, \tag{0.6.26}$$

where

$$\begin{cases} t_k = (\Delta t)k, \quad k = (0, 1, 2, 3, \ldots), \\ x(t) \to x_k \approx x(t_k), \\ h = \Delta t. \end{cases} \tag{0.6.27}$$

Solving for X_{k+1} gives

$$x_{k+1} = (1 - \lambda h)x_k, x_0 \text{ given.} \tag{0.6.28}$$

Inspection of Equation (0.6.25) indicates that all its solutions are positive and decrease to zero monotonically, i.e.,

$$x(t) = x_0 e^{-\lambda t}. \tag{0.6.29}$$

Now by iteration, the solution to Equation (0.6.28) is

$$x_k = x_0(1 - \lambda h)^k, k = (1, 2, 3, \ldots). \tag{0.6.30}$$

However, x_k will only have non-negative, monotonic decreasing solutions if

$$0 < 1 - \lambda h < 1 \Rightarrow h < \frac{1}{\lambda}. \tag{0.6.31}$$

Question: Is it possible to reformulate this forward-Euler scheme to have the correct properties for x_k for all values of the time step-size, h?

One way to proceed is to make the following observation

$$1 - \lambda h = e^{-\lambda h} + O(\lambda^2 h^2), \tag{0.6.32}$$

which can then be rewritten to the form

$$h = \frac{1 - e^{-\lambda h}}{\lambda} + O(\lambda h^2). \tag{0.6.33}$$

The "trick" (or great idea) is to replace h, in Equation (0.6.26) by this expression, i.e,

$$h \to \frac{1 - e^{-\lambda h}}{\lambda} \equiv \phi(\lambda, h). \tag{0.6.34}$$

Note that $\phi(\lambda, h)$, the **denominator function,** has the following properties:

$$\text{(i)} \quad \phi(\lambda, h) = h + O(\lambda h^2). \tag{0.6.35}$$

$$\text{(ii)} \quad \text{Lim}_{\lambda \to 0} \phi(\lambda, h) = h. \tag{0.6.36}$$

$$\text{(iii)} \quad \text{Lim}_{h \to \infty} \phi(\lambda, h) = \frac{1}{\lambda} \tag{0.6.37}$$

$$\text{(iv)} \quad \frac{d\phi(\lambda, h)}{dh} = e^{-\lambda h} > 0. \tag{0.6.38}$$

Thus, for fixed $\lambda > 0$, $\phi(\lambda, h)$ increases from zero at $h = 0$, monotonically to a maximum value of $(1/\lambda)$. Observe that $(1/\lambda)$ is the characteristic time scale of the decaying system.

The new scheme is

$$\frac{x_{k+1} - x_k}{\phi} = -\lambda x_k, \tag{0.6.39}$$

which if solved for x_{k+1}, produces

$$x_{k+1} = e^{-\lambda h} x_k, x_0 \text{ given}. \tag{0.6.40}$$

The corresponding solution is

$$\begin{aligned} x_k &= x_o e^{-\lambda h k}, \quad k = (1, 2, 3, \ldots) \\ &= x_o e^{-\lambda t_k}. \end{aligned} \tag{0.6.41}$$

Observe that

$$x_k = x(t_k), \tag{0.6.42}$$

where x_k is the solution to the forward-Euler discretization scheme and $x(t_k)$ is the discrete-time exact solution to the original differential equation. This result holds for all non-negative values of the time step-size and can be called an **exact finite difference scheme.** It may be that replacing h by a more complex denominator function can allow for the construction of better discretezations schemes for the numerical integration of other differential equations.

0.6.3.3 Modified decay equation

The regular decay differential equation can be modified to take the form

$$\frac{dx}{dt} = -\lambda_1 x - \lambda_2 x^{\frac{1}{3}}, x(0) = x_0 > 0, \tag{0.6.43}$$

where both λ_1, and λ_2 are non-negative parameters.

Inspection of this differential equation (which in fact can be solved exactly, since it is a Bernoulli type equation) indicates that it has $x(t) = 0$ as an equilibrium solutions and this solution is stable.

Consider the value of x, call it $\bar{x}$, for which the magnitude of the two terms on the right-side of Equation (0.6.43) are equal, i.e.,

$$\lambda_1 \bar{x} = \lambda_2 \bar{x}^{\frac{1}{3}}, \tag{0.6.44}$$

then

$$\bar{x} = \left(\frac{\lambda_2}{\lambda_1}\right)^{\frac{3}{2}}. \tag{0.6.45}$$

Small and large values of x are defined as

$$\begin{cases} x \quad \text{small: } 0 \leq x << \bar{x}, \\ x \quad \text{large: } x >> \bar{x}. \end{cases} \tag{0.6.46}$$

Thus, the modified decay equation has the two approximate forms

$$\begin{cases} x \text{ small: } & \dfrac{dx}{dt} \simeq -\lambda_2 x^{\frac{1}{3}}, \\ x \text{ large: } & \dfrac{dx}{dt} \simeq -\lambda_1 x. \end{cases} \tag{0.6.47}$$

The x large approximation holds at the beginning of the decay process and is modelled by the second of Equations (0.6.47). However, for long times, when $x(t)$ is small, the first equation holds.

A very important fact can be derived from the solution to the first differential equation of Equation (0.6.47). This solution is

$$x(t) \approx \left[C - \left(\frac{2\lambda_2}{3}\right)t\right]^{3/2}, t \text{ large}, \tag{0.6.48}$$

where C is positive and

$$t_{large} >> \frac{1}{\lambda}. \tag{0.6.49}$$

Observe that $x(t)$ is not defined for

$$t > \frac{3C}{2\lambda_2}, \tag{0.6.50}$$

because the solution then becomes complex in value. Consequently, from these arguments, we may conclude that the solution to Equation (0.6.43) is zero for some t value, t^*, where

$$t^* = O\left(\frac{3C}{2\lambda_2}\right). \tag{0.6.51}$$

Thus, the solution $x(t)$, of the differential equation modelling this system, namely, Equation (0.6.43) is finite, monotonic decreasing, over the interval $0 < t \leq t^*$, and is zero for $t > t^*$. In other words, the decay process takes a finite time.

We will return to this problem in more detail in a later chapter.

0.7 USE OF HANDBOOKS

Anyone involved in mathematical modelling or scientific research, especially involving differential equations, requires a thorough and reliable guide to the essential features of these equations. In particular, a handbook providing ready access to information from a broad range of mathematical topics is needed. Currently, there exists a rather large set of mathematical handbooks which to varying degrees give helpful definitions, statements of important theorems, relevant formulas, plots of functions, sometimes tables of function values, and often references to the original research literature.

There are mathematical handbooks on almost any conceivable topic in mathematics. They range from listings of "named equations" to giving brief summaries for their origins. Some present and briefly discuss techniques for constructing approximations to solutions and the canditions for which we expect these techniques to be valid. Others show detailed properties of the solutions to well-known differential equations and give discussions on how other equations may be transformed into them.

My three favorite handbooks are

- I.S. Gradshteyn and I.M. Ryzhik, **Table of Integrals, Series, and Products** (Academic Press, New York, 1965).
- A. Jeffrey and H.-H. Dai, **Handbook of Mathematical Formulas and Integrals, 4th Edition** (Elsevier, Amsterdom, 2008).
- D. Zwillinger, **Handbook of Differential Equations** (Academic Press, New York, 1989).

If you own (which you should do) or have easy access to these books, then essentially all of your handbook needs will be satisfied.

The book by Gradshteyn and Ryzhik is the oldest of the three and contains a broader range of mathematical topics than the book by Jeffrey and Dai. Essentially every known indefinite and definite integral is included. Later editions give additional integrals. Of importance is the inclusion of many integrals of the special and related functions. The book of Jeffrey and Dai is in some ways a junior version of the one of Gradshteyn and Ryzhik. However, their book is less information dense and easier to use to locate specific items of interest.

The handbook of Zwillinger is critical for anyone doing mathematical modelling with differential equations. It devotes only one or two pages to each topic, but provides references to the research literature where often background details, including proofs are provided. Another advantage is that this book covers many of the nonlinear differential equations which often (or not) appear during the modelling process.

One, among many valueable aspects of Zwillinger's book, is the inclusion of a section on the "look up technique" for locating exact solutions to both ordinary and partial differential equations. This method, in general, only works for equations having certain forms, but, these specific forms occur frequently in applications. Thus, if a differential equation can be transformed into a known form, then details on the solutions can be found by reading the right references or examining a helpful handbook.

The "look up technique" is presented in Section 33 of the Zwillinger book and covers about 22 pages with an additional nine pages of references. After giving the name of the differential equation, its mathematical representation is given, followed by references as to where the particular equation has been solved or analyzed. The listed equations include ordinary, partial, and systems of such equations. Once a mathematical model has been constructed that is based on differential equations, Zwillinger handbook, section 33, might to the place to go and see if the model contains any of the equations listed in the book.

One final issue needs to be discussed. While all books have something similar to a "Preface," few readers actually read them. This should not be the case. A "Preface" contains important information on the intent of the author, their wishes as to how the book should be used, and the purposes achieved by the writing of the book. A reader's knowledge of these realizations will almostly certainly modify their experience and views on using the book. Also, the author may include their desired procedures for using the book. Pay attention to this. Finally, sometimes the author will devote several pages to state and clarify the symbols used, the notation presented, and the normalization of functions. The latter issue can change or modify the numerical values of coefficients appearing in various formulae in the handbook. So, be careful.

0.8 THE USE OF WIKIPEDIA

In additon to handbooks, another valuable resource is the internet, in particular, searches using wikipedia. Since wikipedia is a free online encyclopedia that can be freely changed and edited by anyone, at any time, it should be used only for an introduction to a topic and not as the final authority. Like most other websites and encyclopedias, it provides general overviews of the subjects it covers, but, as just stated, it should be a place to begin your search to locate information found in standard, reliable, (hopefully) peer-reviewed publications. For a given wikipedia article, the references contained in it are generally more reliable than the article itself and they can be used as guides to additional information. Be aware that wikipedia articles are tertiary sources at best, i.e., they take their information from other primary and secondary sources. Further, they are not references to be cited in formal publicaitons.

In summary, wikipedia can be a very useful source of background information in the early stages of trying to understand some subject or concept. It is an easy and direct procedure for obtaining preliminary information. Also, there exist techniques to aid in becoming a faster, more effective searcher of information on online information sources. One good site is University of London Online Library-Advanced Search Techniques.

Finally, my personal experience has been that on online sites, those dealing with mathematical concepts, special funcitons, and associated plots of these functions, are generally of high level and accurate. Further, the attached "see also," "Notes", "Further reading", and "External links," are both appropriate and very helpful.

0.9 DISCUSSION

After the discussions presented in the previous eight sections, this end this chapter with several remarks:

(i) In general, mathematical modelling is a process that applies mathematics to represent, analyze, make predictions, and provide insights into the phenomena occuring in the natural world.

(ii) A mathematical model is always incomplete. No such model can incorporate every aspect of a system. Therefore, all models are approximations, and simplificaitons and idealizations of the actual system.

(iii) A model should contain the critical features of a system. But, this requires that the modeler acquires a deep and fundamental understanding of the "working" of the system, along with what issues need to be resolved.

(iv) Mathematical, as well as other types of models, are not unique. They can always be improved upon.

(v) The main goals of mathematical models are to explain known results and predict new results.

(vi) The stucture of a good mathematical model often hints at or provides clues to how an improved or better model can be constructed.

(vii) A good model should be consistent, i.e., it should not contain contradictory aspects. However, one should be careful with this requirements, as the Bohratom model illusrates.

(viii) Finally, there are several requirements that are important to being a good modeler. They include

- Having the skills to recognize critical patterns in both the original system and the model;
- Being able to think logically and reason rationally;
- Understanding the original system or issue, and the principles which constraint its behavior;
- Having on appreciation and knowlege of a broad range of mathematical topics.

PROLEMS

Section 0.3

1) Use the methodolgy of Section 0.3.4 to construct/evaluate the sum

$$\frac{1}{3} + \frac{1}{9} + \frac{1}{27} + \cdots .$$

Section 0.4

2) What are, if any, the values of heuristic proofs?

3) Since not all "knowledge" is scientific, then can they ever "conflict" with each? Give possible examples.

Section 0.5

4) Dimensionlize the following differential equations

$$\frac{d^2x}{dt^2} + \omega^2 x^3 = a(1 - b|x|)\left(\frac{dx}{dt}\right)^{\frac{1}{3}}, \tag{0.9.1}$$

$$\frac{\partial u}{\partial t} + au\frac{\partial u}{\partial x} = D\frac{\partial^2 u}{\partial x^2} + \lambda_1\sqrt{u} - \lambda_2 u^2. \tag{0.9.2}$$

Section 0.6

5) Determine the two exact linear independent functions for the differential equation

$$\frac{d^2x}{dt^2} + t\frac{dx}{dt} + x = 0. \tag{0.9.3}$$

Note that you already know that one solution is

$$x_1(t) = \exp\left(-\frac{t^2}{2}\right). \tag{0.9.4}$$

Further, from the elementary theory of linear second-order differential equations, if we know one solution, then the second solution can be determined, generally, in an integral form. Also, observe that

$$\frac{d^2x}{dt^2} + t\frac{dx}{dt} + x = \frac{d}{dt}\left(\frac{dx}{dt} + tx\right) = 0.$$

Section 0.8

Using Wikipedia as a starting point learn about the mathematical properties of the Gamma and Riemann Zeta functions

0.10 NOTES AND REFERENCES

Section 0.1: A concise, but good, presentation of the concept of dynamic consistency is given in

1. R. E. Mickews, Dynamic consistency: a fundamental principle for constructing NSFD schemes for differential equations, **Journal of Difference Equations and Applications** II (2005), 645–653.

Section 0.2: There is a vast literature on the subject of mathematical modelling. We suggest the reader search the internet and examine some of the articles, books, blogs, etc, that currently exist.

Section 0.4: This section is based on the article

2. R. E. Mickens and C. Patterson, What is Science? **Georgia Journal of Science** 24 (2016), No. 2, Article 3 (5 pages).

The following are references from this article which are also referenced in this section

3. K. Devlin, **Mathematics: The Science of Patterns** (Scientific American Library, New York, 1994).

4. L. Flammer, What Science IS (www.indiana.edu/ ensiweb/lessons/unt.s.is.html).

5. J. M. Ziman, **Public Knowledge** (Cambridge University Press, New York, 1968).

A recent book discussing the issues raised in this section is

6. M. Strevens, **The knowledge Machine: How Irrationality Created Modern Science** (Liveright, New York, 2020).

The book puts forth the argument that all differences in scientific interpretations can only be resolved by use of empirical testing.

Section 0.4.1: This section is a compilation of various statements and comments on modelling, science, mathematics, and some related topics. Also, included are several "fun" comments on general issues related to us as human beings. Most of these items are well-known and, as a consequence, no references are given as to who might have stated them, except for a few individuals. In any case, the nature of this book precludes doing this. I hope you enjoy reading these statements, for some provide really insights into the concepts they briefly discuss.

Section 0.5: Two books giving good discussions of both scaling and forming dimensionless variables are

7. H. E. Huntley, **Dimensional Analysis** (Dover, New York, 1947).

8. T. Szirtes, **Applied Dimensional Analysis and Modelling** (McGraw-Hill, New York, 1998).

Section 0.6: Dominant balance, is discussed along with a detailed illustration of several applications, in the books

9. D. Zwillinger, **Handbook of Differential Equations** (Academic Press, Boston, 1989).

See Section 108, pps. 393–395. Also, see the Note to this section for the distinction between "apparent consistency" and "genuine consistency" of the solutions found using this technique.

10. R. E. Mickens, **Mathematical Methods for the Natural and Engineering Sciences, Second Edition** (World Scientific, London, 2017). See Section 13.10, pps. 561–575.

Section 0.7: The advantages of a handbook, if it is free of major errors, is the presentation of a broad range of mathematical formulas, functions and their relationships, tables of numerical values for particular functions, and a summary of various techniques such as how to calculate Fourier series, Laplace transforms, Taylor series, and the construction of partial fraction from rational expressions. When approaching a handbook for the first time, one should examine in detail both its table of contents and index. This should then be followed with a look through the handbook to determine its format, style, and layout. At this point, read the preface to obtain the author's views on how the handbook should be used as a source of mathematical information. Next, pick a topic in which you are familiar and read to see how the handbook deals with it. Follow this by selecting a subject unknown to you and see how it is treated and if you have gained some understanding of it. Finally, every so often repeat these exercises. Doing these things will allow you to dicover what is in the handbook, while at the same time enhance your ability to use it for your needs.

CHAPTER 1

What Is the $\sqrt{N}$?

1.1 INTRODUCTION

Suppose we have a real, positive number, N, and we wish to have the value of its square-root, denoted by $\sqrt{N}$, of course there are two values associated with this quantity, one positive, the second negative. However, our focus will only be on the positive square-root since the negative one differs only by a sign.

The purpose of this chapter is to construct three algorithms for carrying out this task. Generally, we would like to construct a procedure with the properties:

(i) Select the number N and **a priori** choose the number of significant places we wish to have for the value of an approximation to its square-root.

(ii) Construct a procedure which will allow $\sqrt{N}$ to be calculated to any finite number of significant places.

(iii) Use this procedure or algorithm to calculate $\sqrt{N}$ to the prior specified accuracy.

In the presentation to come, we will illustrate our algorithms by use of the specific number, $N = 10$. However, it should be clear that the general methodology holds for any $N > 0$. Also, while our demonstrations are done by hand, anyone with minimum knowledge of computer programming could easily convert these procedures into short programs for implementation on a digital computer.

It needs to be stated that even for the case of N being an integer, the $\sqrt{N}$ is generally an irrational number. This means that only rational approximations to it can be calculated.

In terms of the language of modelling, the "system" here is a question: Given N, a positive, real number, what is $\sqrt{N}$? The mathematical model resolving this question corresponds to the particular algorithm used to calculate its approximate value.

In the next section, we introduce an elementary method to estimate $\sqrt{N}$. We will call it the iterative guessing method (IGM). Section 1.3 provides a technique based on the use of a series expansion, and it will be denoted the series expansion method (SEM). Finally, in Section 1.4, we use a procedure based on Newton's method. It has the advantage of converging to $\sqrt{N}$ very fast from a practical computational point of view.

DOI: 10.1201/9781003178972-1

1.2 INTERATIVE GUESSING

We demonstrate this method by first applying it to a model problem, namely, approximating the $\sqrt{10}$, to five significant places.

Since, $9 < 10 < 16$, it following that

$$3 < \sqrt{10} < 4. \tag{1.2.1}$$

Let us now guess that a first good approximation for $\sqrt{10}$ is

$$X_0 = 3.2, \tag{1.2.2}$$

where

$$X^2 = 10, \tag{1.2.3}$$

where X is the exact value. But

$$X_0^2 = 10.24, \tag{1.2.4}$$

which is larger than 10. This means that our next estimate or guess should be smaller than 3.2, So, take this value, X_1, to be

$$X_1 = 3.1, \tag{1.2.5}$$

which upon squaring gives

$$X_1^2 = 9.61, \tag{1.2.6}$$

a value less than 10. Therefore, the third estimate must be such that

$$3.1 < X_2 < 3.2. \tag{1.2.7}$$

A reasonable selection is

$$X_2 = 3.15, \tag{1.2.8}$$

which gives

$$X_2^2 = 9.9225, \tag{1.2.9}$$

which is somewhat less than 10.

The following Table1.1 gives results for ten levels of guessing.

Inspection of Table 1.1 shows that

$$3.1422 < \sqrt{10} < 3.1623, \tag{1.2.10}$$

and to five significant places the $\sqrt{10}$ is

$$\sqrt{10} = 3.1622\ldots. \tag{1.2.11}$$

This value is to be compared to the more accurate value

$$\sqrt{10} = 3.162277660\ldots. \tag{1.2.12}$$

Note that while this method is some what computational intensive, the computations are elementary, i.e., the multiplication of a number by itself. Further, this procedure can be used to find approximations to any finite order of accuracy for any given positive number N. Finally, note that when the string of digitals in the approximations remain the same from one level to the next, then they may be taken to be the actual significant figures of $\sqrt{N}$.

Table 1.1: Calculation/estimation of $\sqrt{10}$ by iterative guessing method.

k	X_k	X_k^2
0	3.2	10.24
1	3.1	9.61
2	3.15	9.9225
3	3.16	9.9856
4	3.161	9.9919
5	3.162	9.9982
6	3.163	10.0046
7	3.1625	10.0014
8	3.1623	10.001
9	3.1622	9.9995

1.3 SERIES EXPANSION METHOD

A second method to calculate $\sqrt{N}$ is to use series expansions.

To proceed, note that for a real, positive number N, we have

$$I^2 + a = N = (I+1)^2 - b, \tag{1.3.1}$$

where

$$\begin{cases} I = \text{positive integer}, \\ (a, b) = \text{real positive numbers.} \end{cases}$$

For example:

$$N = 31.62 \Rightarrow I = 5, \;\; a = 6.62, \;\; b = -4.38$$

or

$$5^2 + 6.62 = 31.62 = 6^2 - 4.38.$$

Note that the taking square-root of the far-right, middle, and far-left terms in Equation (1.3.1) allows them to be rewritten as follows:

$$\begin{aligned} \sqrt{N} &= \left[I^2 + a\right]^{\frac{1}{2}} \\ &= \left[I^2\left(1 + \frac{a}{I^2}\right)\right]^{\frac{1}{2}} \\ &= I[1+u]^{\frac{1}{2}}, u = \frac{a}{I^2}, \end{aligned} \tag{1.3.2}$$

$$\begin{aligned} \sqrt{N} = I\Bigg[1 + \left(\frac{1}{2}\right)u - \left(\frac{1}{8}\right)u^2 + \left(\frac{1}{16}\right)u^3 \\ - \left(\frac{5}{128}\right)u^4 + \left(\frac{7}{256}\right)u^5 + \cdots\Bigg], \end{aligned}$$

Table 1.2: Calculation/estimation of $\sqrt{10}$ by series expansions for $a = 1$ and $b = 6$.

k	$S_k(1)$	$S_k(6)$
0	3.000000	4.000000
1	3.166666	3.250000
2	3.162036	3.245605
3	3.162294	3.242516
4	3.162276	...

and

$$\sqrt{N} = \left[(I+1)^2 - b\right]^{\frac{1}{2}} \tag{1.3.3}$$

$$= (I+1)^2\left[1 - \frac{b}{(I+1)^2}\right]^{\frac{1}{2}}$$

$$= (I+1)[1-v]^{\frac{1}{2}}, v = \frac{b}{(I+1)^2},$$

$$\sqrt{N} = (I+1)\left[1 - \left(\frac{1}{2}\right)v - \left(\frac{1}{8}\right)v^2 - \left(\frac{1}{16}\right)v^3 - \left(\frac{5}{128}\right)v^4 - \left(\frac{7}{256}\right)v^5 + \cdots\right] \cdot (1.3.3)$$

From the way that (u, v) are constructed, it easy to see that

$$0 < u < 1, \quad 0 < v < 1. \tag{1.3.4}$$

Let $S_k(a)$ and $S_k(b)$ denote the sum of the first $(k+1)$ terms, respectively, on the right-sides of Equations (1.3.2) and (1.3.3). Table 1.2 gives the values of $S_k(a)$ and $S_k(b)$ for $\sqrt{10}$, where $I = 3$, $a = 1$, and $b = 6$. Observe that the $10 = 9 + 1$ expansion converges to the correct five significant value much faster than the $10 = 16 - 6$ series representations, which is converging very slowly.

To repeat or restate from our comments in the previous section, we are using a practical interpretation of the concept of the convergence of a series: When a string of digits in the calculated value of $\sqrt{N}$ does not change in going from one partial sum of the series to the next, which has the addition of one extra term, we take the digits in the sum to be non-changing or fixed for the remainder of the computation. As an illustration of this examine the second column in Table 1.2.

1.4 NEWTON METHOD ALGORITHM

A third procedure to calculate very accurate estimates of the square-root of a positive number is based on the use of Newton's method. Consider a functional relationship between x and y,

$$F(x, y) = 0, \tag{1.4.1}$$

where $F(x,y)$ is given, x is specified, and we wish to determine y iteratively, that is construct a relationship

$$y_{k+1} = \Phi(x, y_k), y_0 = \text{given} \tag{1.4.2}$$

where y_k is the k-th approximation to the solution y of Equation (1.4.1). In general, the procedure goes as follows:

(i) Guess or estimate a starting value y_0.

(ii) Calculate $y_1 = \Phi(x, y_0)$.

(iii) Calculate $y_2 = \Phi(x, y)$.

(iv) Calculate up to a value $k = \bar{k}$, such that

$$\left|y_{\bar{k}+1} - y_{\bar{k}}\right| < \epsilon, \tag{1.4.3}$$

where ϵ is error tolerance or level of numerical precision required for the number y.

The Newton method selects $\Phi(x, y_k)$ to be

$$\begin{aligned} y_{k+1} &\equiv \Phi(x, y_k) \\ &= y_k - \frac{F(x, y_k)}{F_y(x, y_k)}, \end{aligned} \tag{1.4.4}$$

where

$$F_y(x, y_k) = \frac{\partial F(x,y)}{\partial y}\bigg|y = y_k. \tag{1.4.5}$$

If $N > 0$, then the two square-roots of N are $(+\sqrt{N})$ and $(-\sqrt{N})$. For example, if $N = 9$, then its two square-roots are $(+3)$ and (-3). Thus, if one square-root is found, then (for a real, positive number) the other is its negative.

The equation

$$F(N, y) \equiv y^2 - N = 0, \tag{1.4.6}$$

has two solutions corresponding to the square-root of N, i.e,

$$y_x^{(1)} = +\sqrt{N}, y^{(2)} = -\sqrt{N}. \tag{1.4.7}$$

The related Newton iteration scheme is

$$y_{k+1} = \left(\frac{1}{2}\right)\left[y_k + \frac{N}{y_k}\right], \tag{1.4.8}$$

since

$$\frac{\partial F}{\partial y} = 2y, \tag{1.4.9}$$

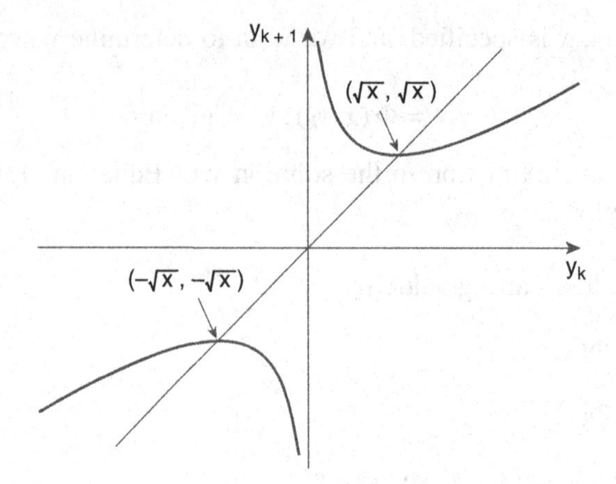

Figure 1.1: Plot of y_{k+1} vs y_k for Equation (1.4.8).

and using this in Equation (1.4.4) gives

$$\begin{aligned} y_{k+1} &= y_k - \frac{\left(y_k^2 - N\right)}{2y_k} \\ &= \left(\frac{1}{2}\right)\left[y_k + \frac{N}{y_k}\right]. \end{aligned} \tag{1.4.10}$$

Examination of the iteration scheme given in Equation (1.4.8) allows the following conclusions to be reached:

(a) y_{k+1} and y_k have the same sign.

(b) $\Phi(N, y_k)$ is an odd function of y_k.

(c) The two fixed points or constant solutions are $(+\sqrt{N})$ and $(-\sqrt{N})$.

(d) For $y_0 > 0$, the iterates y_k converge to $(+\sqrt{N})$, while for $y_0 < 0$, the iterates y_k converge to $(-\sqrt{N})$, i.e.,

$$\int\limits_{k\to\infty} y_k = \begin{cases} y_0 > 0, +\sqrt{N}; \\ y_0 < 0, -\sqrt{N}. \end{cases} \tag{1.4.11}$$

See Figure 1.1 for a plot of y_{k+1} vs y_k, i.e., Equation (1.4.8).

Figures 1.2 and 1.3 show sketches of the initial paths in the (y_k, y_{k+1}) iteration space, respectively, for y_0 positive and $y_0 > \sqrt{N}$ and $y_0 < \sqrt{N}$. For the case where y_0 is negative, the mirror images are obtained.

Table 1.3 provides values of y_k for $N = 10$. The second column uses an initial estimate of $y_0 = 3.2$, while the third column is based on $y_0 = 20$. For $y_0 = 3.2$, the value of $\sqrt{10}$ to five significant places is achieved after just two iterations. Likewise, the same result follows after six iterations for $y_0 = 20$. In any case, a detailed analysis of this scheme shows that whatever $y_0 > 0$ is used the result as $k \to \infty$ converges to the exact value of $\sqrt{10}$.

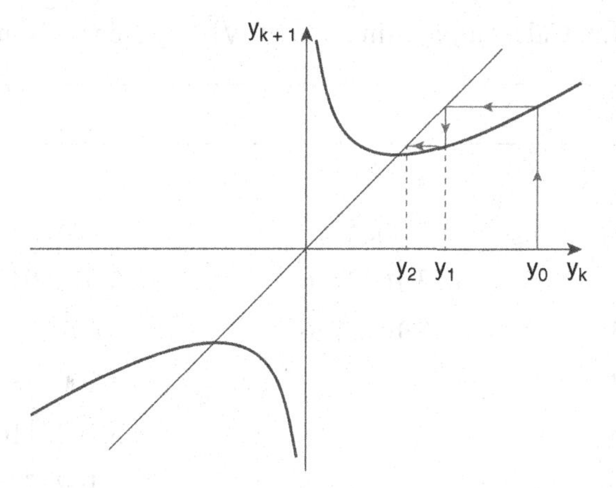

Figure 1.2: Sketch of the initial path in the (y_k, y_{k+1}) space for $y_0 > \sqrt{N}$.

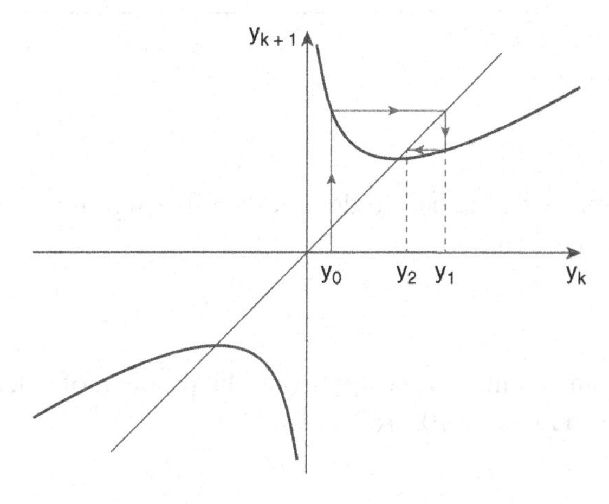

Figure 1.3: Sketch of the initial iteration path in the (y_k, y_{k+1}) space for $O < y_0 < \sqrt{N}$.

1.5 DISCUSSION

This chapter has presented three different methods for calculating approximations to the square-root of a positive number. The most elementary is based on guessing an initial estimate for $\sqrt{N}$ and squaring it for comparison with N. This process is continued until the specified accuracy is obtained. This method may become boring but is very easy to implement.

The second method centers on rewriting N such that an expansion of $\sqrt{1 + x}$ is required to calculate $\sqrt{N}$. Again, this procedure is relatively elementary to carry out, but can be time consuming.

The third method involves the initial use of Newton's method to construct on iteration scheme to calculate the estimates. In general, this method converges very fast and only two or three iterations may be needed to obtain four or five significant figures in the estimate for $\sqrt{N}$, provided y_0 is close to the actual value of $\sqrt{N}$.

While the existence of hand calculators mostly eliminates the necessity of calculating $\sqrt{N}$ by hand, you never really know when such primative techniques may be needed. Now you know of three methods for obtaining accurate estimates of $\sqrt{N}$.

Table 1.3: Calculation/estimation of $\sqrt{10}$ by Newton's method.

K	y_k	y_k
0	3.2	20
1	3.1625	10.25
2	3.1622776	5.6128045
3	3.1622776	3.697225
4	...	3.2009781
5	...	3.1625116
6	...	3.1622776

$\sqrt{10} = 3.162277660\ldots$

PROBLEMS

Section 1.2

1) Use the iterative guessing method to determine a five significant place approximation to the square root of 0.216.

Section 1.3

2) Can the series expansion method be applied to the problem of calculating a five significant place approximation to $\sqrt{0.216}$? Try it.

Section 1.4

3) Calculate $\sqrt{0.216}$, to five significant places, using the Newton method based algorithm.

4) What happens if one uses the Newton method-based algorithm for a negative value of N, i.e., $N < 0$? Try this for $N = -10$. Hint: Use the algorithm

$$y_{k+1} = \left(\frac{1}{2}\right)\left[y_k - \frac{10}{y_k}\right],$$

and plot y_k vs k for $K = 0, 1, 2, \ldots, 100$.

5) Consider the determination of the m-th root of N, where $N > 0$. For this case, we have

$$F(x, y) \equiv y^m - N = 0.$$

Construct the corresponding Newton-based iteration scheme. In particular, consider the cases where $m = \frac{1}{3}$ and $m = -1$.

NOTES AND REFERENCES

Section 1.2: The method of iterative guessing for getting square-roots was taught to me by my grandfather (James Williamson) when I was in the sixth grade. Its a very obvious technique to both understand and use.

Section 1.3: In my own teaching and research, I seldom use the series method, except to obtain first or second order approximations for the expression $(1 + x)^p$, $|x| << 1$ and p is real, i.e.,

$$(1 + x)^p = 1 + px + \left[\frac{p(p-i)}{2}\right] x^2 + O\left(x^3\right).$$

Section 1.4: For additional insights into the properties and applications of Newton's method, and its rate of convergence, see the following two books:

(1) C.M. Bender and S.A. Orszay. **Advanced Mathematical Methods for Scientists and Engineers** (Springer, New York,1999). See pps. 234-235.

(2) R.E. Mickens, **Difference Equations: Theory and Applications, Second Edition** (Chapman and Hall, New York, 1990). See Section 7.2.5.

Note that for N negative, i.e., $N = -|N|$, and

$$y_{k+i} = \left(\frac{1}{2}\right)\left(y_k - \frac{|N|}{y_k}\right), \tag{1.5.1}$$

the selection of a real y_0 gives a real y_k. If then follows that y_k can never approach or converge to the actual value of $i\sqrt{|N|}$, where $i = \sqrt{-1}$. To achieve insight into this situation, the reader should look at pages 237–239 of the book by Bender and Orszay. Their work studies the special case for $N = -3$.

CHAPTER 2

Damping/Dissipative Forces

2.1 INTRODUCTION

Damping and or dissipative forces play very important roles in the modelling and analysis of many systems in the natural and engineering sciences. Dissipation is generally the term used for situations where a conservative system, i.e., a system where the total mechanical energy is constant, is acted upon by friction and/or viscous forces. Damping generally refers to systems which undergo oscillatory behavior. However, for certain cases, no clear distinctions can be made. Within the context of this chapter, we will combine these into the expression damping/dissipative forces, which we denote as DDF.

The main goal of this chapter is to examine the behavior or dynamics of a particle of mass, m_0, moving in the positive x-direction, acted upon by a purely DDF. This DDF is taken to depend only on a power of the velocity with the value of this power being a non-negative number. We find that the common expectation of the particle stopping in a finite distance may not hold and we examine the reasons for this.

Mathematically, Newton's force equation takes the following form for our system

$$m_0 \frac{d^2x}{dt^2} = -F\left(\frac{dx}{dt}\right), \tag{2.1.1}$$

where the velocity is $v(t) = dx(t)/dt$ and $F(v)$ is the DDF. Note that this equation can be rewritten to the form

$$m_0 \frac{dv}{dt} = -F(v). \tag{2.1.2}$$

Also, the initial conditions will be taken as

$$x(0) = 0, \quad \frac{dx(0)}{dt} = V(0) > 0. \tag{2.1.3}$$

This means that our "particle" begins its motion at the origin, with velocity $v_0 = V(0) > 0$, i.e., it is moving in the direction of increasing x values. See Figure 2.1 and Figure 2.2.

In the next section, we state the restrictions which are placed on $F(v)$ such that it is a valid and physically correct DDF. Section 2.3 uses dimensional analysis to construct

DOI: 10.1201/9781003178972-2

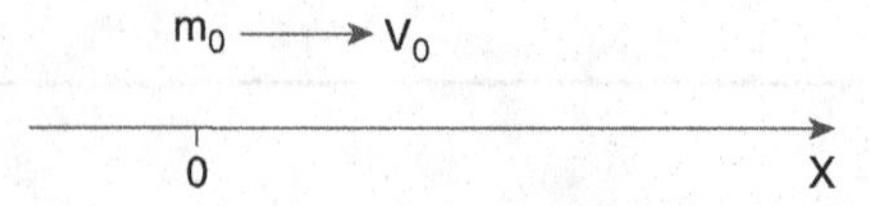

Figure 2.1: A particle of mass m_0, at $t = 0$, with position at $x = 0$ and velocity $V(0) = v_0 > 0$.

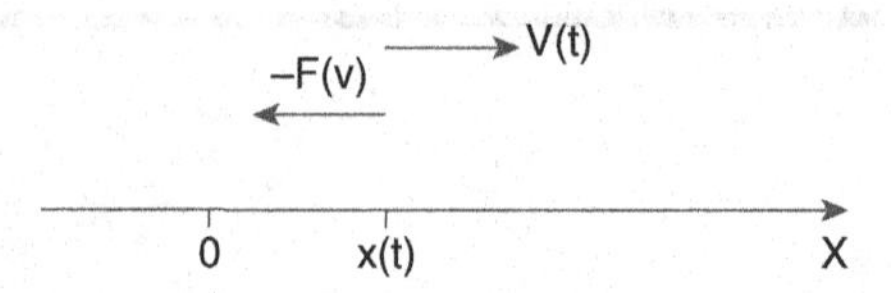

Figure 2.2: At $t > 0$, the particle is located at $x(t)$ and moves with velocity $v(t)$. The only force acting on it is the DDF, $-F(v)$.

two of the most common forms of $F(v)$, namely, the cases where $F(v)$ is either linear or quadratic functions of V. In Section 2.4, we take $F(v)$ to be a single power-law function of V and calculate $x(t)$ and $v(t)$ for six cases of interest. Next a detailed study is made of two, two-term power-law DDF functions. In spite of the fact that the differential equations are nonlinear, their exact solutions can be determined. Finally, in Section 2.6, we summary our results, discuss their significance for modelling, and provide examples of more complex DDF's.

2.2 PROPERTIES OF DDF FUNCTIONS

Replace $F(v)$, in Equation (2.1.2) by $f(v)$,
where

$$F(v) = m_0 f(v), \tag{2.2.1}$$

and obtain

$$\frac{dv}{dt} = -f(v), \tag{2.2.2}$$

with the initial conditions

$$X(0) = 0, \qquad V(0) = v_0 > 0. \tag{2.2.3}$$

The $m_0 f(v)$ is the negative of the DDF.

The following restrictions are placed on the mathematical properties of $f(v)$:

(i) $f(v)$ is continuous in V.

(ii) $f(-v) = -f(v)$.

(iii) $f(v) > 0$, if $v > 0$.

The first condition is a reasonable mathematical requirement and is consistent with the physical data. Condition (ii) is based on the fact that DDF's change sign if the direct of the motion is changed to the opposite motion. Conditions (i) and (ii) together imply that $f(0) = 0$, i.e., there is no DDF when the particle is not in motion. Finally, requirement (iii) is what is expected of any DDF, i.e., the force always acts in a direction opposite to the direction of the velocity.

Another characterization of $f(v)$ is to consider its impact on the kinetic energy of the particle. For the situation under investigation, its Kinetic energy

$$KE \equiv \frac{1}{2} m_0 V^2 \tag{2.2.4}$$

may be taken to be its total energy

$$E(x, v) = KE = \frac{1}{2} m_0 V^2. \tag{2.2.5}$$

If Equation (2.1.2) is written in system form

$$\frac{dx}{dt} = v, \quad \frac{dv}{dt} = -f(v), \tag{2.2.6}$$

then

$$\begin{aligned} \frac{dE}{dt} &= \frac{\partial E}{\partial x}\frac{dx}{dt} + \frac{\partial E}{\partial v}\frac{dv}{dt} \\ &= (0)\frac{dx}{dt} + (m_0 v)[-f(v)], \end{aligned} \tag{2.2.7}$$

and

$$\frac{dE}{dt} = -m_0 v\, f(v) \leq 0. \tag{2.2.8}$$

Since $KE \geq 0$ and its derivative, $dE/dt \leq 0$, then the Kinetic energy continuously decreases to zero as $t \to \infty$.

The class of DDF functions of interest is that characterized by two parameters, α and λ, i.e.,

$$f(v) = \lambda\,[\text{sgn}\,(v)]\ |v|^{\alpha}, \tag{2.2.9}$$

where

$$\lambda > 0, \quad \alpha \geq 0, \tag{2.2.10}$$

and

$$\text{sgn}(v) = \begin{cases} +1, & v > 0, \\ 0, & v = 0, \\ -1, & v < 0. \end{cases} \tag{2.2.11}$$

Inspection of Equation (2.2.9) shows that it satisfies the above stated conditions (i), (ii), and (iii).

For the system described above of a particle of mass m_0, which for $t = 0$ is at the origin, with initial velocity $v(0) = v_0 > 0$, the sgn (v) does not need to be indicated since the velocity is always non-negative and never changes direction. Therefore, the equation of motion, with initial conditions, is

$$\frac{dv(t)}{dt} = -\lambda\, v^{(\alpha)}, \quad x(0) = 0, \quad v(0) = v_0 > 0. \tag{2.2.12}$$

2.3 DIMENSIONAL ANALYSIS AND DDF FUNCTIONS

From deep physical knowledge of fluids and general scientific insightfullness, it is possible to discover the variables upon which the DDFs may depend. Consequently, the application of dimensional analysis will allow the derivation of possible functional forms for the DDFs. This section provides derivation of two such formulas.

2.3.1 $F(v)$ Quadratic in v

Assume that $F(v)$ depends on three variables, (ρ, A, V), where

ρ = density of fluid,
A = cross section of particle, perpendicular to the motion,
V = velocity in the direction of motion.

Based on this information, we assume that $F(v)$ has the form

$$F_1(v) = D_1\, \rho^a\, A^b\, V^c, \tag{2.3.1}$$

where (a, b, c) are to be determined and D_1 is a dimensionless constant or pure number. In terms of (M, L, T) → (mass, length, time), these variables have the following physical units

$$\begin{cases} [\rho] = \dfrac{M}{L^3}, \quad [A] = L^2, \quad [V] = \dfrac{L}{T}. \\ \\ [F_1] = \dfrac{ML}{T^2}, \quad [D_1] = M^0 L^0 T^0. \end{cases} \tag{2.3.2}$$

Placing these results in Equation (2.3.1) gives

$$\begin{aligned} \frac{ML}{T^2} &= \left(\frac{M}{L^3}\right)^a \left(L^{2b}\right)\left(\frac{L}{T}\right)^c \\ &= M^a\, L^{(-3a+2b+c)}\, T^{(-c)}. \end{aligned} \tag{2.3.3}$$

Comparing powers of similar units on the left- and right-sides gives

$$a = 1, \ -3a + 2b + c = 1, \ c = 2, \tag{2.3.4}$$

or

$$a = 1, \ b = 1, \ c = 2. \tag{2.3.5}$$

Therefore, we have

$$F_1(v) = D_1\, \rho\, A\, V^2. \tag{2.3.6}$$

Inspection of Equation (2.3.6) allows the following conclusions to be reached:

(a) The denser the fluid the larger the DDF. For example, a ball moving in water will experience a much larger DDF than an identical ball, having all other conditions the same, moving in the air.

(b) The DDF is proportional to the area perpendicular to the motion. This can be easily demonstrated by taking two sheets of notebook paper and crumbling one into a ball. If both objects are dropped simultaneously from the same height above a floor, the crumbled sheet will hit the floor first.

(c) This DDF depends on the square of the velocity.

2.3.2 $F(v)$ Linear in v

For this case, assume that $F(v)$ depends on the following three variables

η = viscosity of the fluid,
A = cross-section of the particle, perpendicular to the motion,
V = velocity in the direction of motion.

The viscosity has the physical units

$$[\eta] = \left[\frac{\text{force} \cdot \text{time}}{\text{area}}\right] = \frac{M}{LT}. \tag{2.3.7}$$

Therefore, from the assumption

$$F_2 = D_2\, \eta^a A^b V^c, \tag{2.3.8}$$

we obtain

$$\begin{aligned} \frac{ML}{T^2} &= \left(\frac{M}{LT}\right)^a \left(L^{2b}\right) \left(\frac{L}{T}\right)^c \\ &= M^a\, L^{(-a+2b+c)}\, T^{(-a-c)}, \end{aligned} \tag{2.3.9}$$

and

$$a = 1, \quad -a + 2b + c = 1, \quad a + c = 2, \tag{2.3.10}$$

which gives

$$a = 1, \quad b = \frac{1}{2}, \quad c = 1. \tag{2.3.11}$$

Therefore, for this set of variables

$$F_2(v) = D_1\, \eta \sqrt{A}\, v. \tag{2.3.12}$$

Examination of this expression for a DDF gives the following results:

(i) This DDF is linear in the viscosity.
Thus, a particle encounters a larger DDF in honey than in water.

(ii) $F_2(v)$ depends or $\sqrt{A}$ or a linear measure of the particles size perpendicular to its motion.

(iii) Likewise, $F_2(v)$ is a linear function of the velocity.

2.4 ONE TERM POWER-LAW DDF FUNCTIONS

For a one term power-law DDF function, the dynamics is determined by solving the following initial value problem (see Equation (2.2.12)),

$$\begin{cases} \dfrac{dv}{dt} = -\lambda\, v^{\alpha}; \;\; \lambda > 0, \; \alpha \geq 0; \\ x\,(0) = 0, \;\; v\,(0) = v_0 > 0. \end{cases} \tag{2.4.1}$$

Inspection of this expression allows the following conclusions to be reached:

(a) $v\,(t) = 0$ is a solution.

(b) For $v_0 > 0,\;\; V\,(t) \geq 0$ for $t > 0$.

(c) The general solution starts at $v_0 > 0$ for $t = 0$ and monotonically decreases as t increases.

(d) Exact solutions can be determined for this initial-value problem using elementary calculus.

The result in (d) can be demonstrated by observing that the differential equation is separable and can be rewritten as

$$\int v^{-\alpha}\, dV = -\,\lambda \int dt. \tag{2.4.2}$$

If these expressions are integrated we obtain

$$\frac{v^{1-\alpha}}{(1-\alpha)} = -\lambda t \; + \mathrm{C}, \tag{2.4.3}$$

where C is a constant of integration. Once $v\,(t)$ is calculated, then $x\,(t)$ may be determined from

$$x\,(t) = \int_0^t v\,(z) dz. \tag{2.4.4}$$

There are six cases to investigate. These cases depend on the value of the parameter α. In particular, the six cases correspond to what happens to $X\,(t)$ and $V\,(t)$ as $t \to \infty$. The six cases are

(A) $\alpha > 2$,

(B) $\alpha = 2$,

(C) $1 < \alpha < 2$,

(D) $\alpha = 1$,

(E) $0 < \alpha < 1$,

(F) $\alpha = 0$.

Case A : $\alpha > 2$

Let $\alpha = 2 + \beta$, where $\beta > 0$, then $V(t)$ is given by the expression

$$v(t) = \frac{v_0}{\left[1 + \lambda(1+\beta)\, v_0^{(1+\beta)}\, t\right]^{\frac{1}{1+\beta}}}. \tag{2.4.5}$$

For large t, $v(t)$ is asymptotic to

$$v(t) \sim \frac{C_1}{t^{\left(\frac{1}{1+\beta}\right)}}, \tag{2.4.6}$$

where C_1 can be written in terms of (λ, β, v_0). This result implies that

$$\operatorname*{Lim}_{t\to\infty} V(t) = 0 \tag{2.4.7}$$

The substitution of Equation (2.4.5) into the intergrand of Equation (2.4.4) will allow $x(t)$ to be calculated. However, our real interest is in $x(\infty)$, i.e.,

$$x(\infty) = \operatorname*{Lim}_{t\to\infty} x(t) = \int_0^\infty v(z)dz. \tag{2.4.8}$$

Since, in Equation (2.4.6),

$$0 < \frac{1}{1+\beta} < 1, \tag{2.4.9}$$

it follows that

$$X(\infty) = \infty, \tag{2.4.10}$$

i.e., $x(t)$ becomes unbounded as $t \to \infty$. This means that while $v(t) \to 0$ as $t \to \infty$, the particle's motion carries it an infinite distance from the starting position at $x = 0$. In other words, **while there is a DDF acting on the particle, it never stops moving**.

Case B : $\alpha = 2$

For this case, $v(t)$ and $x(t)$ are, respectively, given by the expressions

$$v(t) = \frac{v_0}{1 + \lambda\, v_0\, t}, \tag{2.4.11}$$

$$x(t) = \left(\frac{1}{\lambda}\right) \operatorname{Ln}(1 + \lambda\, v_0\, t) \tag{2.4.12}$$

Therefore,

$$\operatorname*{Lim}_{t\to\infty} v(t) = 0, \operatorname*{Lim}_{t\to\infty} x(t) = \infty. \tag{2.4.13}$$

The conclusions regarding the general properties of $v(t)$ and $x(t)$ are the same as for Case A.

Case C: $1 < \alpha < 2$

Let $\alpha = 1 + \gamma$, where $0 < \gamma < 1$. The functions $v(t)$ and $x(t)$ are, respectively,

$$v(t) = \frac{v_0}{\left[1 + \lambda\gamma\, v_0^{\gamma}\, t\right]^{\left(\frac{1}{\gamma}\right)}}, \tag{2.4.14}$$

$$x(t) = \left[\frac{v_0^{(1-\gamma)}}{\lambda(1-\gamma)}\right]\left\{1 - \frac{1}{\left(1 + \lambda\gamma\, v_0^{\gamma}\, t\right)^{\left(\frac{1-\gamma}{\gamma}\right)}}\right\}. \tag{2.4.15}$$

Also,

$$\lim_{t\to\infty} v(t) = v(\infty) = 0, \tag{2.4.16}$$

$$\lim_{t\to\infty} x(t) = x(\infty) = \frac{v_0^{(1-\gamma)}}{\lambda(1-\gamma)} < \infty. \tag{2.4.17}$$

For this case, $V(\infty) = 0$, while $x(\infty)$ has a finite value.

Case D : $\alpha = 1$

The solutions for $v(t)$ and $x(t)$ are

$$v(t) = v_0\, e^{-\lambda t}, \quad x(t) = \left(\frac{v_0}{\lambda}\right)\left(1 - e^{-\lambda t}\right), \tag{2.4.18}$$

and

$$\lim_{t\to\infty} v(t) = 0, \quad \lim_{t\to\infty} x(t) = \frac{v_0}{\lambda}, \tag{2.4.19}$$

Case E : $0 < \alpha < 1$

Let $\alpha = 1 - \delta$, where $0 < \delta < 1$. For this situation

$$v(t) = \left[v_0^{\delta} - \delta\lambda t\right]^{\frac{1}{\delta}}. \tag{2.4.20}$$

Observe that $v(t)$ becomes complex valued for $t > t^*$, where

$$t^* = \frac{v_0^{\delta}}{\delta\,\lambda}. \tag{2.4.21}$$

Since $v(t) = 0$ is a solution and since only real valued solutions for this problem have physical meaning, we conclude that $v(t)$ is actually the following piecewise, continuous function

$$v(t) = \begin{cases} \left[v_0^{\delta} - \lambda\,\delta\, t\right]^{\frac{1}{\delta}}, 0 < t \le t^*, \\ 0, \ t > t^*. \end{cases} \tag{2.4.22}$$

Likewise,

$$x(t) = \begin{cases} \left[\dfrac{1}{\lambda\,(0+\delta)}\right]\left\{v_0^{1+\delta} - \left[v_0^{\delta} - \lambda\delta t\right]^{\frac{1+\delta}{\delta}}\right\}, \\ \qquad \text{for } 0 < t \le t^*; \\ \dfrac{v_0^{1+\delta}}{\lambda(1+\delta)}, \quad t > t^*. \end{cases} \tag{2.4.23}$$

Thus, the particle stops in a finite time, t^*, and only travels a finite time.

Case F: $\alpha = 0$

The equation of motion for this case is

$$\frac{dv}{dt} = -\lambda, \; v(0) = v_0 > 0, \; x(0) = 0, \tag{2.4.24}$$

and the solution is

$$v(t) = v_0 - \lambda\, t. \tag{2.4.25}$$

However, this function becomes negative for $t > \bar{t}$, where

$$\bar{t} = \frac{v_0}{\lambda}. \tag{2.4.26}$$

This situation is unphysical, i.e., it is not expected that the particle will slow to a stop and then reverse direction and speed up. Consequently, the correct physical solution is

$$v(t) = \begin{cases} v_0 - \lambda t, & 0 < t \le \bar{t}, \\ 0, \text{ if } t > \bar{t}. \end{cases} \tag{2.4.27}$$

Using

$$x(t) = \int_0^t v(z)\, dz, \tag{2.4.28}$$

we find

$$x(t) = \begin{cases} v_0 t - \left(\frac{\lambda}{2}\right) t^2, \; 0 < t \le \bar{t}, \\ \frac{v_0^z}{z\lambda}, \; t > \bar{t}. \end{cases} \tag{2.4.29}$$

Thus, the particle travels only a finite distance and it does this in a finite time $\bar{t}$.

Summary

For all values of non-negative α, the velocity decreases monotonic to zero. However, for $\alpha > 1$, it takes an unlimited amount of time to do so. For $0 \le \alpha < 1$, the time for the particle to stop is finite.

With regard to $x(t)$, for $\alpha > 2$, the particle never stops moving. While its velocity is always decreasing, the final location of the particle is infinitely far from its starting location.

For $1 \le \alpha < 2$, the particle stops in a finite distance, but takes an infinite time to do so.

In the interval, $0 \le \alpha < 1$, both the displacement, $x(t)$, and the velocity, $v(t)$, show finite-time dynamics. This means that there exists a time, t_1, such that the motion ceases for times larger than t_1.

2.5 TWO TERM POWER-LAW DDF FUNCTIONS

We now consider DDF functions consisting of two different DDF's, i.e., they take the form

$$F(v) = -A\ V^p - B\ V^q, \tag{2.5.1}$$

where

$$A > 0,\ B > 0, p > 0,\ q > 0. \tag{2.5.2}$$

For the case where $p = 1$, i.e.,

$$\frac{dv}{dt} = -A\ v - B\ v^q, \tag{2.5.3}$$

this first-order, nonlinear differential equation is a Bernoulli type equation and can be solved for its exact solution.

In the following two sub sections, we examine the cases, respectively, for $q = 2$ and $q = \frac{1}{2}$.

2.5.1 $p = 1$ and $q = 2$

For these values of the power-law parameters, we have

$$\frac{dv}{dt} = -\lambda_1 v - \lambda_2 v^2. \tag{2.5.4}$$

This equation can be linearized by means of the transformation

$$v(t) = u(t)^{-1}. \tag{2.5.5}$$

Substituting this into Equation (2.5.4) and using

$$\frac{dv}{dt} = (-)\left(\frac{1}{u^2}\right)\frac{du}{dt}, \tag{2.5.6}$$

gives

$$-\left(\frac{1}{u^2}\right)\frac{du}{dt} = -\left(\frac{\lambda_1}{u}\right) - \left(\frac{\lambda_2}{u^2}\right), \tag{2.5.7}$$

which reduces to

$$\frac{du}{dt} = \lambda_1 u + \lambda_2, \tag{2.5.8}$$

a first-order, linear differential equation. It general solution is

$$u(t) = u_H(t) + u_P(t), \tag{2.5.9}$$

where $u_H(t)$ and $u_P(t)$ are the, respective, homogeneous and particular solutions, given by the expressions

$$u_{(H)}(t) = A\ e^{\lambda_1 t}, \quad u_P(t) = -\left(\frac{\lambda_2}{\lambda_1}\right), \tag{2.5.10}$$

where A is an arbitrary integration constant.

Therefore,

$$u(t) = A\, e^{\lambda_1 t} - \left(\frac{\lambda_2}{\lambda_1}\right), \tag{2.5.11}$$

and A can be determined from the condition

$$\begin{aligned} u(0) &= \frac{1}{v_0} \\ &= A - \left(\frac{\lambda_2}{\lambda_1}\right). \end{aligned} \tag{2.5.12}$$

Therefore, A is

$$\begin{aligned} A &= \frac{1}{v_0} + \frac{\lambda_2}{\lambda_1} \\ &= \frac{\lambda_2\, v_0\, + \lambda_1}{\lambda_1\, v_0}, \end{aligned} \tag{2.5.13}$$

and

$$u(t) = \left(\frac{\lambda_2\, v_0 + \lambda_1}{\lambda_1\, v_0}\right) e^{\lambda_1 t} - \left(\frac{\lambda_2}{\lambda_1}\right). \tag{2.5.14}$$

Finally, using $v(t) = 1/u(t)$, we obtain

$$v(t) = \frac{1}{\left(\frac{\lambda_2\, v_0 + \lambda_1}{\lambda_1\, v_0}\right) e^{\lambda_1 t} - \left(\frac{\lambda_2}{\lambda_1}\right)}. \tag{2.5.15}$$

Since,

$$e^{\lambda_1 t} > 1 \text{ for } t \geq 0, \frac{\lambda_2\, v_0 + \lambda_1}{\lambda_1\, v_0} > \frac{\lambda_2}{\lambda_1}, \tag{2.5.16}$$

the denominator on the right-side of the expression in Equation (2.5.15) is never zero and $v(t)$ is defined for $t \geq 0$.

The following is a listing of some of the important properties of $v(t)$:

(a) Taking $t \to 0$, in Equation (2.5.15) gives

$$\lim_{t \to 0} v(t) = v_0. \tag{2.5.17}$$

(b)

$$\lim_{t \to \infty} v(t) = 0. \tag{2.5.18}$$

(c) $v(t)$, for large t, asymptotic to the expression

$$v(t) \sim \left(\frac{\lambda_1\, v_0}{\lambda_2\, v_0 + \lambda_1}\right) e^{-\lambda_1 t}. \tag{2.5.19}$$

(d) For "small" $v(t)$, the solution is dominated by the term $(-\lambda_1\, v)$ in Equation (2.5.4). This fact is consistent with the principle of dominant balance, which informs us that for small v, $v^2 << V$.

(e) In summary, $v(t)$ decreases monotonic t from v_0, at $t = 0$, to $V(\infty) = 0$, at $t = \infty$

Given $v(t)$, $x(t)$ can be calculated from the relation

$$x(t) = \int_0^t v(z)\, dz. \tag{2.5.20}$$

We leave it as a homework project to evaluate $x(t)$ and examine it properties as a function of t.

2.5.2 $p = 1$ and $q = \frac{1}{2}$

For these values of the parameters, Equation (2.5.1) is

$$\frac{dv}{dt} = -\lambda_1 v - \lambda_3 v^{\frac{1}{2}}, \quad v(0) = v_0 > 0. \tag{2.5.21}$$

Comment: If we were considering general motion along the two directions on the x-axis, then the second term would be written

$$\lambda_3\, v^{\frac{1}{2}} \rightarrow \lambda_3 \left[\operatorname{sgn}(v)\right]\, |v|^{\frac{1}{2}}. \tag{2.5.22}$$

This also applies to the $\lambda_2\, V^2$ term in the previous subsection; it would be rewritten as

$$\lambda_2\, v^2 \rightarrow \lambda_2 \left[\operatorname{sgn}(v)\right]\, v^2. \tag{2.5.23}$$

The transformation

$$u = \sqrt{v} \quad \text{or } v = u^2, \tag{2.5.24}$$

gives for Equation (2.5.20),

$$\frac{du}{dt} = -\bar{\lambda}_1 u - \bar{\lambda}_3, \quad u(0) = u_0 = \sqrt{v_0} \tag{2.5.25}$$

where

$$\lambda_1 = 2\bar{\lambda}_1, \ \lambda_2 = 2\bar{\lambda}_2. \tag{2.5.26}$$

Solving for $u(t)$ and enforcing the initial condition gives

$$u(t) = \left(\sqrt{v_0} + \frac{\lambda_3}{\lambda_1}\right) e^{-\bar{\lambda}_1 t} - \left(\frac{\lambda_3}{\lambda_1}\right), \tag{2.5.27}$$

and for $v(t)$ the result,

$$V(t) = \left[\left(\sqrt{v_0} + \frac{\lambda_3}{\lambda_1}\right) e^{-\bar{\lambda}_1 t} - \left(\frac{\lambda_3}{\lambda_1}\right)\right]^2. \tag{2.5.28}$$

The requirement that $u(t) \geq 0$ implies that there exists a time, t^*, such that $u(t^*) = 0$. Thus, we have

$$\left(\sqrt{v_0} + \frac{\lambda_3}{\lambda_1}\right) e^{-\bar{\lambda}_1 t^*} = \frac{\lambda_3}{\lambda_1}, \tag{2.5.29}$$

and the function $u(t)$ must be defined as

$$u(t) = \begin{cases} \left[\left(\sqrt{v_0} + \frac{\lambda_3}{\lambda_1}\right) e^{-\bar{\lambda}_1 t} - \left(\frac{\lambda_3}{\lambda_1}\right)\right], 0 \le t \le t^*, \\ 0, \text{ for } t > t^*. \end{cases} \quad (2.5.30)$$

The corresponding function form for $v(t)$ is

$$v(t) = \begin{cases} \left[\left(\sqrt{v_0} + \frac{\lambda_3}{\lambda_1}\right) e^{-\bar{\lambda}_1 t} - \left(\frac{\lambda_3}{\lambda_1}\right)\right]^2, 0 \le t \le t^*, \\ 0, \text{ for } t > t^*. \end{cases} \quad (2.5.31)$$

The following is a summary of the properties of the solution to Equation (2.5.20):

(i) Taking $t = 0$, gives

$$V(0) = v_0. \quad (2.5.32)$$

(ii) For $0 \le t \le t^*$, $v(t)$ decreases monotonic from $V(0) = v_0 > 0$ to $V(t^*) = 0$.

(iii) For "small" $v(t)$, the solution is dominated by the term $\left(-\lambda_3 \sqrt{v}\right)$ in Equation (2.5.20) which gives rise to finite-time dynamics

We leave it as a homework project to calculate $x(t)$ and evaluate it properties.

2.6 DISCUSSION

This chapter has been focused on the dynamics of DDF (dissipative, damping forces) acting on a particle, with initial velocity v_0, moving in one dimension. We found that the particle's velocity decreases monotonic to zero. In particular, for a single power-law DDF, i.e.,

$$\begin{aligned} \frac{dv}{dt} &= -F(v) \\ &= -\lambda v^{\alpha}, \end{aligned} \quad (2.6.1)$$

finite-time dynamics occurs if $0 \le \alpha < 1$, i.e., the particle comes to a complete stop in a finite time. It was also determined that for $\alpha \ge 2$, the particle does not stop in at a finite distance from its starting point, i.e., the particle ceases moving only after it has travelled an infinite distance. Based on our experiences in the actual physical universe, the following conclusion can be reached:

One term power-law DDFs should not have $\alpha \geq 2$.

For general, integer-value, power-law DDFs, i.e., DDFs having the form

$$F(v) = \text{sgn}(v) \sum_{K=1}^{N} f_K |v|^{K}, \quad (2.6.2)$$

where N is a finite integer and

$$F_K \ge 0, \quad K = (1, 2, 3, \ldots), \quad (2.6.3)$$

the same restriction should apply.

Note that the addition of just the single term

$$F_\alpha(v) = \lambda\,[\text{sgn}(v)]\,|v|^\alpha, 0 < \alpha < 1, \tag{2.6.4}$$

to Equation (2.6.2) i.e.,

$$\overline{F}(v) = \lambda\,[\text{sgn}(v)]\,|v|^\alpha + \sum_{K=1}^{N} f_K|v|^K, \tag{2.6.5}$$

forces the dynamics to stop in a finite-time. This result is a consequence of the principle of dominant balance applied to the differential equation

$$\frac{dv}{dt} = -\overline{F}(v). \tag{2.6.6}$$

As a final issue, consider an alternative representation of the parameter α in Equation (2.6.4). The current expression requires the use of the sign-function to insure that $F(v)$ is an odd function of V. However, for $0 < \alpha < 1$, the following more restricted in possible values, form can be used

$$\begin{cases} \alpha \to p(n,m) = \dfrac{2n+1}{2m+1}, \\ (n,m) : 0, 1, 2, ...;\ n < m. \end{cases} \tag{2.6.7}$$

Under these conditions, $p(n,m)$ is a proper fraction and is bounded above by the value one, i.e.,

$$0 < p(m,n) < 1. \tag{2.6.8}$$

Also, $p(m,n)$ is odd in the sense that

$$(-x)^{p(m,n)} = (-1)\,x^{p(m,n)}. \tag{2.6.9}$$

Note that if $q(m,n)$ is taken to be

$$\begin{aligned} q(m,n) &= 1 - p(m,n) \\ &= \frac{2(m-n)}{2m+1}, \end{aligned} \tag{2.6.10}$$

then $q(m,n)$ is even, i.e.,

$$(-x)^{q(m,n)} = x^{q(m,n)}. \tag{2.6.11}$$

Within this framework, the "natural value" of P is $p(1,0) = 1/3$. This corresponds to the smallest possible values of m and n, namely, $n = 0$ and $m = 1$.

PROBLEMS

Section 2.2

1. Explain the physical basis for each of the mathematical properties stated for $f(v)$.

2. Suppose an even function of v is selected as a DDF function. What if any difficulties arise?

Section 2.4

3. Is it possible to have an actual physical DDF such that it is proportional to a negative power of the velocity? Are any physical laws violated? Consider, for example,

$$F(v) = \frac{A}{v^{1/3}}, A > 0.$$

Section 2.6

4. Physical measurements, at any given times, are always limited to a finite number of significant digits. Relate this fact to the possible significance of using fractional values for the powers of the velocity in DDF representations.

NOTES AND REFERENCES

Section 2.2: For a single particle, of mass m, Newton's force law is

$$m\frac{d^2x}{dt^2} = F\left(x, \frac{dx}{dt}\right),$$

where the force can dependent on both the location and velocity of the particle.

Section 2.3: The linear drag force is usually called Stokes' law. Its derivation is given in

1. G.G. Stokes, On the effects of the internal friction of fluids on the motion of pendulums, **Transactions of the Cambridge Philosophical Society**, Vol. 9 (1850), 8-106.

This equation appears on page 51.

The quadratic in velocity law is generally attributed to Lord Rayleigh (John William Strutt). Stokes' law holds for low speeds where smooth or laminar flow occurs, while the quadratic on Lord Rayleigh relation is valid at higher speeds where the flow is turbulent. For a general discussion of this phenomena, see

2. G.K. Batchelor, **Introduction to Fluid Dynamics** (Cambridge University Press, Cambridge, 1967).

Section 2.4: This section based on the following paper

3. R.E. Mickens and K. Oyedeji, Investigation of power-law damping/dissipative forces, are Xiv: 1405. 4062 v1 [physics, comp-ph] 16 May 2014.

Section 2.5: The two differential equations studied in this section are particular cases of the Bernoulli equation.

$$\frac{dy}{dx} + P(x)y = Q(x)y^n, n \neq 0, 1.$$

A listing of nine special first-order differential equations and their solutions is given in

4. M.R. Spiegel, Advanced Mathematics for Engineers and Scientists (McGraw-Hill, Schaum's Outline Series, New York, 1999). See pps. 39-41.

Section 2.6: A general discussion of DDFs possibly leading to finite-time dynamics for 1-dim, nonlinear oscillators is given in the manuscript.

5. R.E. Mickens and K. Oyedeji, Damping/dissipative forces leading to finite-time dynamics (Clark Atlanta University, Department of Physics; Atlanta, GA 30314, USA; June 2012).

CHAPTER 3

The Thomas-Fermi Equation

3.1 INTRODUCTION

One of the most famous and important differential equations in mathematical physics is the Thomas-Fermi equation (TFE),

$$\frac{d^2y}{dx^2} = \frac{y^{3/2}}{\sqrt{x}}, \quad y(0) = 1\,, \quad y(\infty) = 0. \tag{3.1.1}$$

Observe that this is a boundary-value problem since $y(x)$ is specified at $x = 0$ and $x = \infty$. Further, inspection indicates that $y(x)$ is singular at $x = 0$ since its second derivative is unbounded, i.e.,

$$\begin{aligned} \left.\frac{d^2y}{dx^2}\right|_{x=0} &= \underset{x\to 0}{\text{Lim}} \frac{y^{3/2}}{\sqrt{x}} \\ &= \underset{x\to 0}{\text{Lim}} \frac{1}{\sqrt{x}} \\ &= +\infty\,, \end{aligned} \tag{3.1.2}$$

where x approaches zero through positive values. This means that while $y(x)$ may have a series expansion at $x = 0$, it cannot be a Taylor series.

Another interesting feature of the TFE is that for physical meaningful solutions where $y(x)$ is real, it follows that the following requirements must hold

$$x \geq 0, \qquad y(x) \geq 0. \tag{3.1.3}$$

Also, the TFE has $y(x) = 0$ as a solution. This fact, combined with the additional restriction that the second derivative of $y(x)$ is positive, i.e.,

$$y''(x) > 0, \quad 0 < x < \infty, \tag{3.1.4}$$

implies that $y(x)$ is a concave upward function. Putting this together with the initial conditions, $y(0) = 1$ and $y(\infty) = 0$, we conclude that

$$y'(x) < 0, \quad 0 \leq x < \infty. \tag{3.1.5}$$

DOI: 10.1201/9781003178972-3

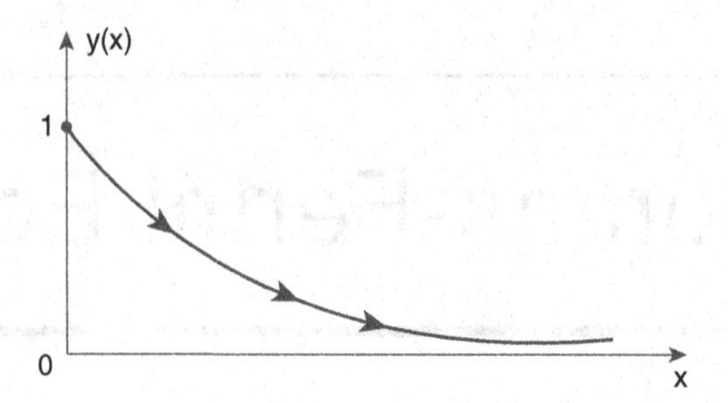

Figure 3.1: General behavior of $y(x)$ vs x.

(This issue will be discussed again in Section 3.2.1). Therefore, $y(x)$ is expected to start with the value one at $x = 0$ and then decrease monotonic to the value zero at $x = \infty$. This general behavior is shown in Figure 3.1.

The TFE is a nonlinear, second-order differential equation for which no exact general solution is known to exist in terms of a finite combination of the elementary functions. As a consequence and because of its importance for atomic physics, hundreds of research papers have been published on the construction of analytical approximations to its solutions. This chapter considers this problem as an issue in mathematical modelling, i.e., the "system" is the TFE, as given in Eq. (3.1.1), with the mathematical modelling consisting of constructing mathematical approximations to the actual solution.

The Notes and References to this section gives a short list of published articles which individually includes almost all of the important previous work on the construction and analysis of approximate solutions to the TFE. The main purpose of the current presentation is to use a new methodology to construct such approximate solutions. This is done with the use of the concept of dynamic consistency and the tight restrictions that it imposes on any a prori selection of an explicit representation of $y(x)$. For our purpose, we consider rational approximations for $y(x)$ and demonstrate that rather elementary rational structures can provide good representations to $y(x)$. A major advantage of this procedure is that there are no free parameters that need to be determined by fits to numerical solutions of Equation (3.1.1). However, in addition to the boundary conditions, given in Equation (3.1.1), we need to know the slope of $y(x)$ at $x = 0$, i.e.,

$$y'(0) = -B, \quad B > 0. \tag{3.1.6}$$

However, B can be computed "to practically arbitrarily high accuracy" and to ten significant places has the value

$$B = 1.588071022\,(6). \tag{3.1.7}$$

The chapter proceeds as follows: Section 3.2 lists and discusses a number of general exact properties of the TFE and its solutions consistent with Equations (3.1.1) and (3.1.6). The concept of dynamic consistency (DC) is introduced in Section 3.3 and we explain its application to the current problem. This principle of DC is then used to directly construct two elementary rational approximations for $y(x)$. We end the chapter with a summary of obtained results and a brief discussion of another problem that has been resolved through application of methods similar to those used for the TFE.

3.2 EXACT RESULTS

One of the remarkable aspects of the TFE is that many of the features of its solutions may be derived in the absence of having an exact explicit representation of its actual solution. We now list several of these results.

(i) One can try to determine if Equation (3.1.1) has special types of exact solution. It is clear that simple elementary solutions, i.e.,

$$y(x) = C_1 e^{-at}, \tag{3.2.1}$$

or even

$$y(x) = C_2 e^{-t^b}, \tag{3.2.2}$$

do not exist. In these expressions, (C_1, C_2, a, b) are constants. Next, the following power law type solution can be assumed,

$$y(x) = AX^{\alpha}, \tag{3.2.3}$$

where A and α are to be determined. Substitution into Equation (3.1.1) gives

$$\alpha(\alpha - 1)\ AX^{\alpha-2} = A^{\frac{3}{2}}\ X^{\frac{3\alpha}{2}-\frac{1}{2}}. \tag{3.2.4}$$

Equating the coefficients and power of X gives

$$\alpha = -3, \quad A = 144, \tag{3.2.5}$$

and

$$y_s(x) = \frac{144}{x^3}. \tag{3.2.6}$$

This is an exact solution of the TFE and contains no arbitrary parameters. Consequently, it is not a special case of the (unknown) general solution. This function is a singular solution of the TFE and is denoted by $y_s(x)$. In fact, it can be demonstrated that for any solution, $y(x)$, it is true that

$$\lim_{x\to\infty}(x^3 y(x)) = 144. \tag{3.2.7}$$

(See E. Hille, 1969.)

(ii) So, how is $y_s(x)$ to be interpreted? The curve $y(x) = y_s(x)$ separates the bounded and unbounded solutions of the TFE, as indicated in Figure 3.2. Note that these two types of solutions are distinguished by the following criteria

$$\begin{cases} \text{bounded solutions:} & y(0)\ \text{finite;} \\ \text{unbounded solutions:} & y(0)\ \text{infinite.} \end{cases} \tag{3.2.8}$$

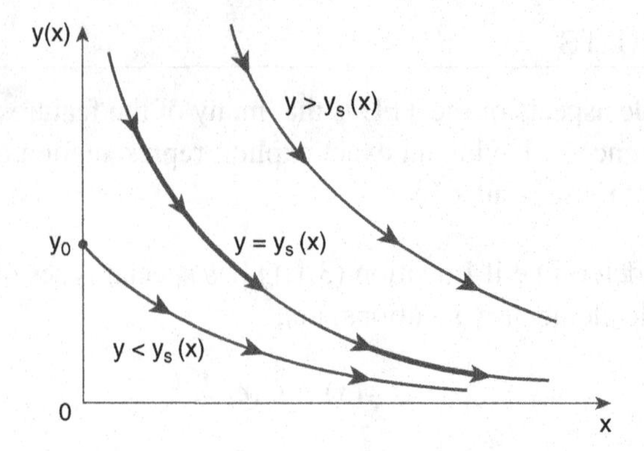

Figure 3.2: The singular solution, $y = y_s(x)$, separates the bounded and unbounded solutions of the TFE.

(iii) The physical phenomena for which the TFE was derived requires that $y(x)$ and $y'(x)$ satisfy the conditions

$$\begin{cases} y(0) = 1, \quad y(\infty) = 0, \\ y(x) > 0, \quad y'(x) < 0, \quad 0 \le x < \infty. \end{cases} \tag{3.2.9}$$

Since $y''(x) > 0$, for $0 < x < \infty$, all solution curves are concave upward, consistent with the general flow of the solution trajectories shown in Figure 3.2.

In a little more detail, we have for the bounded solutions

$$\begin{cases} 0 < y(x) < y(0), \\ -B = y'(0) < y'(x) < 0, \\ \qquad\qquad y''(x) > 0, \\ \text{for } 0 < x < \infty. \end{cases} \tag{3.2.10}$$

In addition,

$$y''(x) = O\left(\frac{1}{\sqrt{x}}\right), \text{ as } x \to 0^+. \tag{3.2.11}$$

(iv) For $y(0) = 1$ and $y'(0) = -B$, then for small x, $y(x)$ has the series representation

$$\begin{aligned} y(x) = &\left[1 - Bx + \left(\frac{1}{3}\right)x^3 - \left(\frac{2B}{15}\right)x^4 + \cdots\right] \\ &+ x^{3/2}\left[\frac{4}{3} - \left(\frac{2B}{5}\right)x + \left(\frac{3B^2}{70}\right)x^2 + \left(\frac{2}{27} + \frac{B^3}{252}\right)x^3 + \cdots\right]. \end{aligned} \tag{3.2.12}$$

Observe that Equation (3.2.12) is not a Taylor series; it is a special case of more general expansions called Puiseux series. We now illustrate how one can obtain such an expansion.

For "small" x, the TFE can be writen as

$$y''(x) \simeq \frac{1}{\sqrt{x}}, \tag{3.2.13}$$

since $y(0) = 1$. Integration once gives

$$y'(x) \simeq \int \frac{dx}{\sqrt{x}} = C_1 + 2x^{\frac{1}{2}}, \tag{3.2.14}$$

and using

$$y'(0) = -B, \tag{3.2.15}$$

we obtain $C_1 = -B_1$, and

$$y'(x) = -B_1 + 2x^{\frac{1}{2}}. \tag{3.2.16}$$

Integrating again and using the condition $y(0) = 1$, gives

$$y(x) \simeq 1 - Bx + \left(\frac{4}{3}\right) x^{3/2}, \tag{3.2.17}$$

which the first three lowest power of x in Equation (3.2.12).

To proceed further and calculation higher power terms, some computational scheme must be constructed to allow us to do so. The following iteration method does this. It is just an extension of what we did above. Let $y_k(x)$ be a sequence of functions defined as the solutions of the second-order differential equations

$$\begin{cases} y''_{k+1}(x) = \frac{[y_k(x)]^{3/2}}{\sqrt{x}}, \quad k = (0, 1, 2, \ldots) \\ y_k(0) = 1, \quad y'_k(0) = -B, \\ y_0(x) = 1. \end{cases} \tag{3.2.18}$$

A simple calculation gives

$$y_1(x) = 1 - Bx + \left(\frac{4}{3}\right) x^{3/2}. \tag{3.2.19}$$

Therefore,

$$\begin{aligned} y''_2(x) &= \frac{[y_1(x)]^{3/2}}{\sqrt{x}} \\ &= \frac{\left[1 - Bx + \left(\frac{4}{3}\right) x^{3/2}\right]^{3/2}}{\sqrt{x}}. \end{aligned} \tag{3.2.20}$$

Expanding the bracket and retaining the first two terms gives

$$\left[1 - Bx + \left(\frac{4}{3}\right)x^{\frac{3}{2}}\right]^{\frac{3}{2}} = 1 - \left(\frac{3B}{2}\right)x + \cdots \tag{3.2.21}$$

and the following expression for $y_2''(x)$

$$y_2''(x) = \frac{1}{\sqrt{x}} - \left(\frac{3B}{2}\right)x. \tag{3.2.22}$$

If this equation is integrated twice, and the two initial conditions are imposed, then we obtain for $y_2(x)$ the expression

$$\begin{aligned} y_2(x) &= 1 - Bx + \left(\frac{4}{3}\right)x^{\frac{3}{2}} - \left(\frac{2B}{5}\right)x^{\frac{5}{2}} \\ &= [1 - Bx] + x^{\frac{2}{3}}\left[\frac{4}{3} - \left(\frac{2B}{5}\right)x\right], \end{aligned} \tag{3.2.23}$$

and they correspond, respectively, to the first two terms in the brackets on the right-side of Equation (3.2.12).

If we continue the calculation to higher-order approximations, then careful attention must be paid to what terms are retained in the expansion of $[y_k(x)]^{-3/2}$. In general, one more term is included as we go from the k to the $(k+1)$ level of the approximation.

(v) A number of "sum rules" can be derived for $y(x)$. A sum rule is an integral relation involving some or all of the functions $y(x)$ and $y'(x)$. Four such sum rules for the TFE are

$$\int_0^\infty \sqrt{x}\; y(x)^{\frac{3}{2}} dx = 1. \tag{3.2.24}$$

$$\int_0^\infty y(x)dx = \left(\frac{1}{2}\right)\int_0^\infty x^{\frac{3}{2}} y(x)^{\frac{3}{2}} dx, \tag{3.2.25}$$

$$\int_0^\infty \left[\left(\frac{dy}{dx}\right)^2 + \frac{y^{\frac{5}{2}}}{\sqrt{x}}\right] dx = B, \tag{3.2.26}$$

$$\int_0^\infty \frac{y(x)^{\frac{3}{2}}}{\sqrt{x}}\, dx = B. \tag{3.2.27}$$

The derivations of these sum rules are not difficult and we illustrate how this can be done by consider Equation (3.2.26).

To begin, multiply the TFE by $y(x)$ and obtain

$$yy'' = \frac{y^{5/2}}{\sqrt{x}}. \tag{3.2.28}$$

Next integrate both sides, apply integration by parts, and evaluate the resulting expressions at $x = 0$ and $x = \infty$. Doing this gives

$$y(\infty)y'(\infty) - y(0)y'(0) - \int_0^\infty \left[y^1(x)\right]^2 dx = \int_0^\infty \frac{y^{\frac{5}{2}}}{\sqrt{x}} dx. \tag{3.2.29}$$

Using the values

$$y(0) = 1, \quad y'(0) = -B, \quad y(\infty) = y'(\infty) = 0, \tag{3.2.30}$$

we arrive at the result in Equation (3.2.26).

3.3 DYNAMIC CONSISTENCY

Dynamic consistency (DC) is a principle that allows us to assess the "closeness" or "fidelity" of two systems based on their mutually shared properties.

Definition

Let A and B be two different systems. Let A have the property P. If B also has the property P, then B is said to be dynamic consistent to A, with respect to P.

Comment

The systems A and B may not be the same type of object or structure. For example, A could be an actual physical systems, while B could be a particular mathematical model of A. Or, A could be a system of differential equations, while B could be particular discretizations of these equations. Consequently, DC is of general applicability and is available for use in a broad range of situations.

In the following section, two proposed approximations to the solution of the TFE are given and analyzed. These ansatzes are constructed such that both are DC with the (unknown) exact solution, $y(x)$, with respect to the following properties:

(i) Initial and boundary conditions

$$y(0) = 1, \quad y'(0) = -B, \quad y(\infty) = 0; \tag{3.3.1}$$

(ii) Boundedness, non-negativity, and monotonicity

$$0 < y(x) \leq 1, \quad -B \leq y'(x) < 0, \quad 0 \leq x < \infty; \tag{3.3.2}$$

(iii) Asymptotic condition

$$\lim_{x\to\infty} \left(x^3 y(x)\right) = 144; \tag{3.3.3}$$

(iv) Agreement with $y(x)$ in the small x limit, i.e.

$$y(x) = 1 - Bx + \left(\frac{4}{3}\right) x^{\frac{3}{2}} + \cdots . \tag{3.3.4}$$

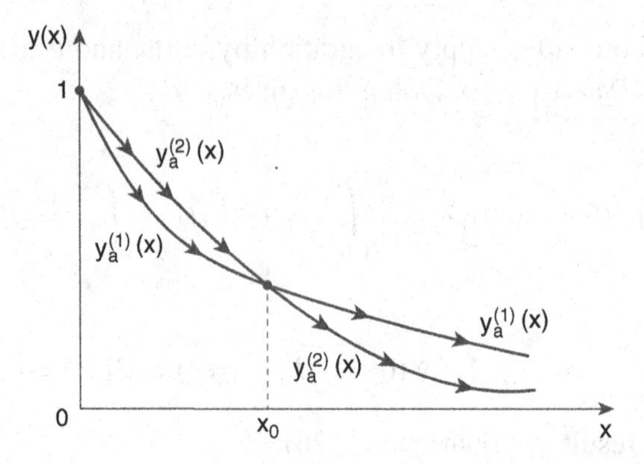

Figure 3.3: Plots of $y_a^{(1)}(x)$ and $y_a^{(2)}(x)$ vs x. The curves intersect at $x_0 \simeq 1$.

Comment

The exact solution to the TFE is unknown. We place a further, but major, restriction on the approximate solutions, $y_a(x)$, by assuming that they are rational functions in the variable $\sqrt{x}$. This assumption is clearly consistent with the small x expansion given in Equation (3.2.12).

3.4 TWO RATIONAL APPROXIMATIONS

The following two ansatzes are selected as approximations to the solution of the TFE:

$$y_a^{(1)}(x) = \frac{1}{1 + Bx + \left(\frac{1}{144}\right) x^3}, \tag{3.4.1}$$

$$y_a^{(2)}(x) = \frac{1}{1 + Bx - \left(\frac{4}{3}\right) x^{\frac{3}{2}} + Cx^2 + \left(\frac{1}{144}\right) x^3} \tag{3.4.2}$$

where

$$C = \frac{B^2}{2}, \tag{3.4.3}$$

with

$$B = 1.588071\ldots. \tag{3.4.4}$$

Comparison of these two representations shows that the above three properties, given in Equations (3.3.1) to (3.3.3), are satisfied and that $y_a^{(1)}(x)$ and $y_a^{(2)}(x)$, respectively, are correct in the first two and three terms of Equation (3.3.4).

Figure 3.3 provides an illustrative representation of the plots of $y_a^{(1)}(x)$ and $y_a^{(2)}(x)$ vs x. While the two curves have the same general shape and lie close to each other, they do not coincide. Further, they intersect at some value, $x = x_0$,

$$\begin{cases} y_a^{(1)}(x) < y_a^{(2)}(x), & 0 \le x < x_0, \\ y_a^{(1)}(x_0) = y_a^{(2)}(x_0), & x = x_0, \\ y_a^{(1)}(x) > y_a^{(2)}(x), & x > x_0, \end{cases} \tag{3.4.5}$$

where $x_0 \simeq 1$.

Table 3.1: Numerical values of $y(x)$ and $y_a(x)$.

x	$y(x)$ numerical*	$y_a^{(1)}(x)$	$y_a^{(2)}(x)$
0.0	1.0000	1.0000	1.0000
0.5	0.6070	0.5571	0.6102
1.0	0.4240	0.3853	0.3964
2.0	0.2430	0.2363	0.1817
4.0	0.1840	0.1283	0.0578
5.0	0.0789	0.1020	0.0378
10.0	0.0243	0.0420	0.0093
25.0	0.0035	0.0067	0.0013
40.0	0.0011	0.0020	0.0005

*P.S. lee and T.-Y. Wu, Chinese Journal of Physics, Vol. 35 (6–11) (1997), 737–741.

Table 3.1 compares the numerics of $y_a^{(1)}(x)$ and $y_a^{(2)}(x)$ with that of a fourth-order Runge-Kutta numerical method for selected values of x in the range 0–40. Both the analytic approximations and the numerical integration results are consistent and in very good qualitative agreement with each other.

Another measure of the accuracy or validity of the above approximate solution constructions is to see how well they satisfy the integral relations (sum rules) given in Section 3.2. In particular, we consider the one given by Equation (3.2.26), i.e.,

$$\int_0^{\infty} \left[\left(\frac{dy}{dx} \right)^2 + \frac{y^{\frac{5}{2}}}{\sqrt{x}} \right] dx = B, \tag{3.4.6}$$

since it has the structure of an "energy" type integral (Taylor, 2005) which appears prominently in classical physics. Inspection of the details of this sum rule suggests the following correspondences

$$\left(\frac{dy}{dx} \right)^2 \to \text{kinetic energy density,}$$

$$\frac{y^{\frac{5}{2}}}{\sqrt{x}} \to \text{potential energy density.}$$

Note that this sum rule also provides an overall consistency check on the methodology used to construct the approximate solutions, $y_a^{(1)}(x)$ and $y_a^{(2)}(x)$, since they both contain the parameter B and Equation (3.4.6) implies that the integral must be equal to B. Thus, the nearness of the integral to B indicates the appropriateness of our method.

Substituting $y_a^{(1)}(x)$ and $y_a^{(2)}(x)$ into the left-side of Equation (3.4.6) and then using numerical integration to evaluate the integrals, the following results are found for the respective integrals

$$B_1 = 1.584744, \quad B_2 = 1.592931 \tag{3.4.7}$$

which is to be compared to the "exact value"

$$B = 1.588071. \tag{3.4.8}$$

The corresponding fractional percentage errors are

$$E_1 = \left(\frac{B - B_1}{B}\right) \cdot 100 = 0.21\%, \tag{3.4.9a}$$

$$E_2 = \left(\frac{B - B_2}{B}\right) \cdot 100 = -0.27\%, \tag{3.4.9b}$$

One conclusion is that the sum rule, stated in Equation (3.4.6), is satisfied to about 0.24% error for both $y_a^{(1)}(x)$ and $y_a^{(2)}(x)$.

3.5 DISCUSSION

The methodology used to construct the ansatzes for $y_a^{(1)}(x)$ and $y_a^{(2)}(x)$ is based on formulating rational approximations for $y(x)$ such that each is consistent with the restrictions given in Section 3.3, i.e., Equations (3.3.1) to (3.3.4). It should be noted that an important feature of this procedure is that the process is essentially algebraic, with a little Taylor series expansion thrown in, and also requires no new parameters to be introduced. In summary, $y_a^{(1)}(x)$ and $y_a^{(2)}(x)$ are determined just from the x-small and x-large behaviors of the exact solution, $y(x)$.

A word of caution should now be given regarding approximations to the solution of the TFE. The TFE is based on a number of both physical and mathematical assumptions. Thus, it is not an exact formulation of the basic physical phenomena for which it is supposed to describe. So the following question arises: What is the significance or meaning of approximate solutions to an approximate differential Equation? The fact that various approximate solutions have been used to provide valuable and "accurate" information on atomic and other physical systems certainly suggests or implies that these approximations to $y(x)$ are not meaningless.

At the level of mathematical modelling the critical aspect of this chapter is the use of the idea of mathematical modelling to help resolve issues related to finding (approximate) solutions to a differential equation. For this case, the system is the differential equation, including initial and boundary conditions, along with exact known properties of the solutions for x-small and large. Thus, mathematical modelling can be used to answer questions involving mathematics itself, especially, for problems aring in applied mathemation.

Finally, we restate, in somewhat different form than previously stated, our definition of a mathematical model.

Definition: Mathematical Model

A **mathematical model** is a description of the **essential features** of a system that uses mathematical language and structures to describe the behavior of the system.

PROBLEMS

Sections 3.1 and 3.2

1) The differential equation

$$y''(z) = y(z)^2 - zy(z)y^1(z),$$

models certain heat transfer phenomena in fluids. The boundary conditions are

$$y^1(0) = -\sqrt{3}, y(\infty) = 0.$$

With the requirement $y(0) > o$, determine the general properties of $y(z)$ and $y^1(z)$ over the interval

$$0 \leq z < \infty.$$

See the article by Mickens and Wilkins (2019).

Sections 3.3 and 3.4

2) Construct a rational approximation to the solution of the equation given in Problem 1. Make sure you state and understand the exact Features of $y(z)$ that are incorporated into the approximate solution.

NOTES AND REFERENCES

Section 3.1: This chapter is based heavily on a lightly rewritten version of the article

1. R.E. Mickens and I.H. Herron, Approximate rational solutions to the Thomas-Fermi equation based on dynamic consistency, **Applied Mathematics Letters**, Vol. 116 (2021), 106994. DOI: 10.1016/j.aml.2020.106994

A similar type of rational approximation to a nonlinear differential equation is presented in the article

2. R.E. Mickens and J. Ernest Wilkins, Jr., Estimation of $y(0)$ for the boundary-value problem: $y'' = y^2 - zyy', y'(0) = -\sqrt{3}, y(\infty) = 0$, **Communications in Applied Analysis**, Vol. 23, No. 1 (2019), 137–146.

The references to the original work of Fermi and Thomas are

3. E. Fermi, Un methodo statistico per in la determinazione di alcunepriorieta dell'atome', **Rend. Acad. Naz. Lincei**, Vol. 6, (1927), 602–607.

4. L.H. Thomas, The calculation of atomic fields, **Proceedings of the Cambridge Philosophical Society**, Vol. 23, (1927), 452–548.

Other articles with references to the research literature on the Thomas-Fermi equations are

5. E.B. Baker, The application of the Thomas-Fermi statistical model to the calculation of potential distribution in positive ions, **Physical Review**, Vol. 16 (1930), 630–647.

6. M.A. Noor, and S.T. Mohyud-Din, Homotopy perturbation method for solving the Thomas-Fermi equation using Pade' approximants, **International Journal of Nonlinear Science**, Vol. 8 (#1) (2009), 27–31.

A good, quick introduction to the Thomas-Fermi equation is

7. Wikipedia: "Thomas-Fermi Equation."

Section 3.2: An article written by a mathematician which discusses some of the properties of the Thomas-Fermi equation is

8. E. Hille, On the Thomas-Fermi equation, **Proceedings of the National Academy of Science USA**, Vol. 62 (#1) (1969), 7–10.

Section 3.3: A good discussion of the concept and application of "dynamic consistency" is given in the article

9. R.E. Mickens, Dynamic consistency: A fundamental principle for constructing nonstandard finite difference schemes for differential equations, **Journal of Difference Equations and Applications**, Vol. 11 (#7) (2005), 645–653.

Section 3.4: For conservative, one-dimension, one particle systems, we have

$$\begin{aligned}\left(\frac{1}{2}\right)m\left(\frac{dx}{dt}\right)^2 + V(x) &= \text{Energy}\\ &= \text{constant},\end{aligned}$$

where

$$\begin{aligned}\text{Kinetic energy} &= \frac{1}{2}m\left(\frac{dx}{dt}\right)^2\\ \text{potential energy} &= V(x).\end{aligned}$$

For more details, see the book

10. J.H. Taylor, **Classical Mechanics** (University Science Books, Sausalito, CA, 2005).

Another article containing a large listing of papers on analytical methods and numerical techniques related to the Thomas-Fermi equation is

11. F. Bayatbabolghani and K. Parand, Using Hermite function for solving Thomas-Fermi equation, **International Journal of Mathematical, Computational Science and Engineering**, Vol. 8 (#1) (2014), (no pages indicated).

A table shows the results of two different numerical calculations of Thomas-Fermi solutions.

CHAPTER 4

Single Population Growth Models

4.1 INTRODUCTION

Populations, however defined, generally change their magnitude as a function of time. The main goals here are to provide some differential equation mathematical models as to how these populations change, construct the corresponding solutions, analyze the properties of these solutions, and indicate some applications. For the case of living biological populations, we assume that all environmental and/or cultural factors operate on a time scale which is much longer than the intrinsic time scale of the population of interest. If this holds true, then the mathematical model takes the following form for a single population.

$$\frac{dP(t)}{dt} = f(P), \quad P(0) = P_0 \geq 0, \tag{4.1.1}$$

where $P(t)$ is the value of the population P at time t. The function $f(P)$ is what distinguishes one model from another. So, how is $f(P)$ to be selected?

First, we expect Equation (4.1.1) to have the structure

$$\begin{aligned} \frac{dP}{dt} &\equiv \text{(growth factors)} - \text{(declining factors)}, \\ &= g(P) - d(P), \end{aligned} \tag{4.1.2}$$

where

$$g(0) = 0 = d(0), \tag{4.1.3}$$

and this implies

$$f(0) = 0. \tag{4.1.4}$$

Note that the requirement of Equation (4.1.4) is related to the "axiom of parenthood." This axiom states "that every organism must have parents..." there is no spontaneous generation of organismsa (Hutchinson, 1978).

DOI: 10.1201/9781003178972-4

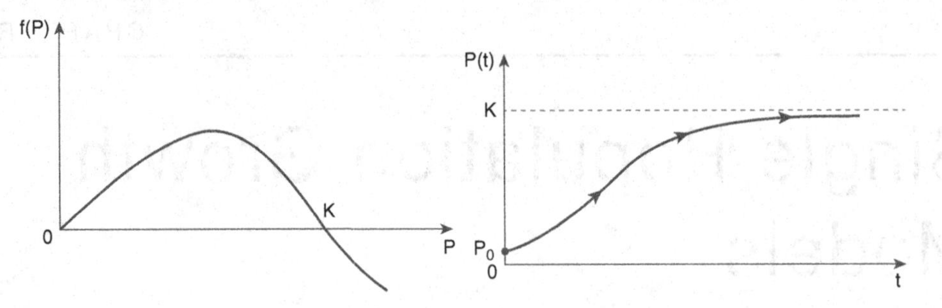

Figure 4.1: (a) Plot of $f(P)$ vs P. (b) General features of the curve $P(t)$ as a function of time.

Second, for all of the models to be examined, the functions $g(P)$ and $d(P)$ will be either power or rational functions, i.e.,

$$g(P)/d(P) = \begin{cases} \text{power: } aP^m, \\ \text{rational: } \dfrac{aP^m}{1 + bP^n}, \end{cases} \tag{4.1.5}$$

where n and m are non-negative numbers which may not be integers, and the parameters a and b are positive numbers.

Since many growing populations start at small values and increase to a larger limiting value, there, we will only examine systems having these features. This implies that $f(P)$ has the following properties:

$$\begin{cases} f(0) = 0, \\ f(K) = 0, \\ f(P) > 0, \quad 0 < P < K, \\ f(P) < 0, \quad P > K. \end{cases} \tag{4.1.6}$$

These general results are illustrated in Figure 4.1.

Several comments need to be made. First, in what follows no derivation, in the mathematical sense, is actually done. We merely write down a number of first-order differential equations for which the $f(P)$s have the properties stated in Equation (4.1.6). Second, for any particular model, the inclusion of other items/variables from it environment, such as available water and food, technological advances, etc., would require introducing other variables, and this would convert the new model into one represented by a system of differential equations. Third, these models only capture several of the gross features of the actual population dynamics. The hope is that this truncation provides an adequate description for our purposes.

This chapter is organized as follows. Section 4.2 discusses the logistic equation, its solutions and their properties, and gives two applications. Section 4.3 examines another logistic type model for growth of a single population, the Gompertz model. It was derived as a simple model for the growth of certain of tumors. Section 4.4 discusses three other single population models which have appeared in the research literature. Section 4.5 introduces the Allee effect and explains the usefulness of this particular modification of the logistic equation. The chapter end with a summary of what has been covered.

4.2 LOGISTIC EQUATION

4.2.1 Introduction

Perhaps the simplest differential equation modelling single population growth is the so-called **logistic equation**. Its applications have been extended to model phenomena for chemical reactions, harvesting of animals such as fish, resource recovery, technological developments, and social and cultural practices. It is based on the assumption that the population under study is well-mixed, i.e., spatially homogeneous.

The logistic differential equation is

$$\frac{dP}{dt} = aP - bP^2; \quad a > 0, b > 0, \quad P_{(0)} = P_0 > 0, \tag{4.2.1}$$

where the first term is interpreted as representing the "births" of population $P(t)$, with the second term giving the death rate of this population. There are two constant or equilibrium solutions, i.e.,

$$\bar{P}^{(1)} = 0, \quad \bar{P}^{(2)} = \frac{a}{b} \equiv K. \tag{4.2.2}$$

In the solution space, $P(t)$ and t, $P(t)$ has the general features shown in Figure 4.1. Starting from an initially small value P_0, the population increases monotonically to the value K. In fact, $\bar{P}^{(2)} = K$ is a stable equilibrium of the system in the sense that

$$\lim_{t \to \infty} P(t) = \bar{P}^{(2)} = K, \quad P_0 > 0. \tag{4.2.3}$$

An important aspect of the logistic differential equation is that its exact solution can be calculated and this is done in the next section.

Note the following curious fact. If Equation (4.2.1) is written as

$$\frac{dP}{dt} = B(P)P, \tag{4.2.4}$$

where $B(P)$ is defined as the **Net birthrate**, then

$$B(P) = a - bP. \tag{4.2.5}$$

This function has the following properties:

(i) $B(P)$ is a decreasing function of P. This implies that the net birthrate decreases with increases of population.

(ii) $B(P)$ is zero at $P = K$, i.e.,

$$B(K) = B(\bar{P}^{(2)}) = 0. \tag{4.2.6}$$

(iii) From (i) and (ii), it follows that

$$B(P): \begin{cases} > 0, & \text{for } 0 \leq P < K, \\ 0, & P = K, \\ < 0, & P > K. \end{cases} \tag{4.2.7}$$

(iv) Finally, we have

$$B(0) = a > 0. \tag{4.2.8}$$

This last result implies that even in the absence of population members, the net birthrate is non-zero and positive. Given the usual interpretation of net birthrate, this makes no sense. A way out of this difficulty is to say that the logistic model should not be interpreted as a net birthrate function multiplied by the population. Further, this conclusion may be extended to other single population growth models.

4.2.2 Solution of Logistic Equation

Equation (4.2.1) is a separable, first-order, nonlinear differential equation and can be rewritten as

$$\begin{aligned} \frac{dP}{dt} &= aP - bP^2 \\ &= bP\left(\frac{a}{b} - P\right) \\ &= bP(K - P), \end{aligned} \tag{4.2.9}$$

or

$$\frac{dP}{P(K-P)} = bdt \tag{4.2.10}$$

Using the method of partial fractions, we have

$$\frac{1}{P(K-P)} = \frac{A}{P} + \frac{B}{K-P}, \tag{4.2.11}$$

where A and B are to be determined. If both sides are multiplied by $P(K-P)$, then the following result is obtained

$$\begin{aligned} 1 &= A(K-P) + BP \\ &= (B-A)P + AK. \end{aligned} \tag{4.2.12}$$

Comparing the coefficients of the constant and linear term in P gives the relations

$$B - A = 0, \; AK = 1, \tag{4.2.13}$$

when solved for A and B yields

$$A = B = \frac{1}{K}. \tag{4.2.14}$$

If these results are substituted into the left-side of Equation (4.2.10) and then integrated, we obtain

$$\left(\frac{1}{K}\right)\int_{P_0}^{P}\left[\frac{1}{Q}+\frac{1}{K-Q}\right]dQ=\int_{0}^{t}bdz, \tag{4.2.15}$$

which upon integrating gives

$$\left(\frac{1}{K}\right)[\mathrm{Ln}\,P-\mathrm{Ln}\,(P-K)]_{t=0}^{t}=bt. \tag{4.2.16}$$

But

$$Kb=\left(\frac{a}{b}\right)b=a, \tag{4.2.17}$$

and

$$\mathrm{Ln}\left(\frac{P}{P-K}\right)-\mathrm{Ln}\left(\frac{P_0}{P_0-K}\right)=at. \tag{4.2.18}$$

Combining the two expressions on the right-side using the formula

$$\mathrm{Ln}\,(f)-\mathrm{Ln}\,(g)=\mathrm{Ln}\left(\frac{f}{g}\right), \tag{4.2.19}$$

and then taking the antilogarithm of the resulting expression, i.e.,

$$X=\mathrm{Ln}\,(y)\Rightarrow y=e^{x}, \tag{4.2.20}$$

and solving for $P(t)$, finally gives

$$P\,(t)=\frac{P_0K}{P_0+(K-P_0)\exp\,(-at)}. \tag{4.2.21}$$

This is the solution to the logistic differential equation as represented in Equation (4.2.1).

We can check the correctness of this expression by looking at its valucs for $t=0$ and $t=\infty$, i.e.,

$$P\,(0)=\frac{P_0K}{P_0+(K-P_0)}=\frac{P_0K}{K}=P_0, \tag{4.2.22}$$

$$P\,(\infty)=\lim_{t\to\infty}\;P\,(t)=\frac{P_0K}{P_0}=K. \tag{4.2.23}$$

Both values are in agreement with our a priori expectations.

4.2.3 Harvesting

There are many single population systems for which harvesting takes place. Fowl or fish farms are examples of such systems. Harvesting is a removal of a certain number of the population during each time period that the harvesting takes place. The most direct way of harvesting is to use a strategy where a constant number, H, of individuals are removed during each time period. For this situation, the logistic differential equation gets modified to the form

$$\frac{dp}{dt}=aP-bP^2-H,\;H\geq 0, \tag{4.2.24}$$

where H is the harvesting rate.

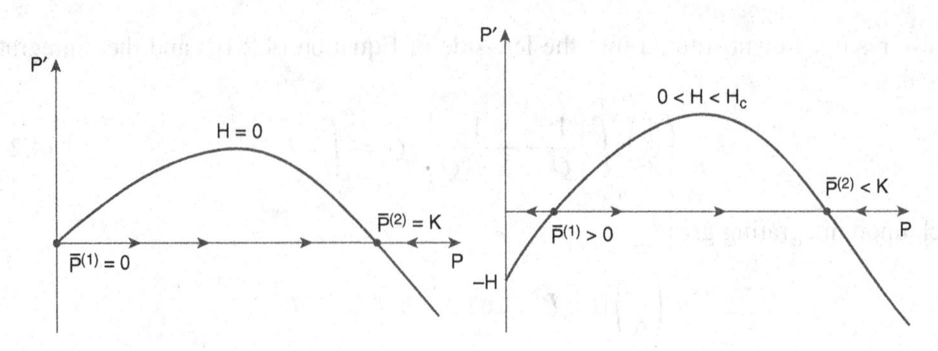

Figure 4.2 (a, b): Plots of $P' = dP/dt$ for various values of the harvesting rate: (a) $H = 0$; (b) $0 < H < H_c$, where $H_c = a^2/4b$.

The equilibrium or constant populations, $\bar{P}$, correspond to $dP/dt = 0$ or

$$-H + a\bar{P} - b\bar{P}^2 = 0. \tag{4.2.25}$$

However, the dynamics of this harvesting system can be easily understood if we examine $P' = dP/dt$ vs P plots. There are four cases to examine as shown in Figure 4.2:

(1) If $H = 0$, then we have the usual logistic equation situation. There is an unstable equilibrium state at $P(t) = \bar{P}^{(1)} = 0$ and a stable equilibrium state at $P(t) = K$. For positive initial values, the final state is located at $P(t) = K$, i.e.,

$$\lim_{t\to\infty} P(t) = K, \; P_0 > 0, \; H = 0. \tag{4.2.26}$$

(2) For $0 < H < H_c$, where $H_c = a^2/4b$, there are two equilibrium states, $\bar{P}^{(1)}$ and $\bar{P}^{(2)}$, satisfying the conditions

$$0 < \bar{P}^{(1)} < \bar{P}^{(2)} < K. \tag{4.2.27}$$

The state at $\bar{P}^{(1)}$ is unstable, while the one at $\bar{P}^{(2)}$ is stable. Consequently, any P_0 in the interval

$$0 \le P_0 < \bar{P}^1, \tag{4.2.28}$$

leads to the total depletion of the population.

(3) If $H = H_c$, there is a single, semi-stable equilibrium state. For a realistic harvesting situation, this corresponds to all $P_0 > 0$ going to zero, i.e., the population is depleted.

(4) For $H > H_c$, the population is depleted.

Figures 4.3, 4.4, and 4.5 give the major features of the solution behaviours for these four cases.

A better model for harvesting is to construct one such that $\bar{P}^{(1)} = 0$ for all harvesting efforts. A simple way to accomplish this is to take the harvesting effort proportional to the population available at the time of harvesting. This gives the model

$$\frac{dP}{dt} = aP - bP^2 - EP, \tag{4.2.29}$$

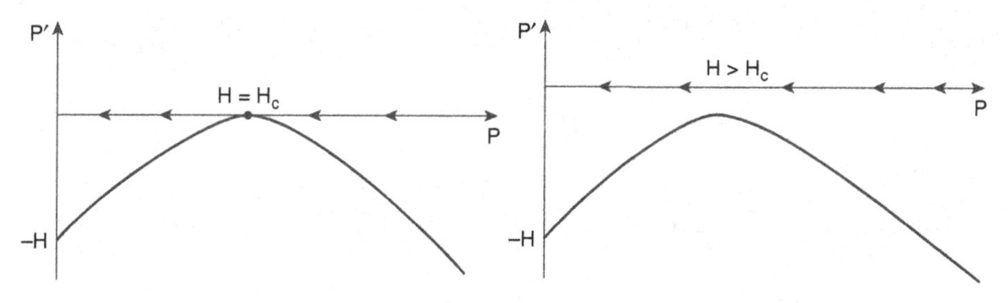

Figure 4.3 (a, b): Plots of $P' = dP/dt$ for (a) $H = H_c$; (b) $H > H_c$, where $H_c = a^2/4b$.

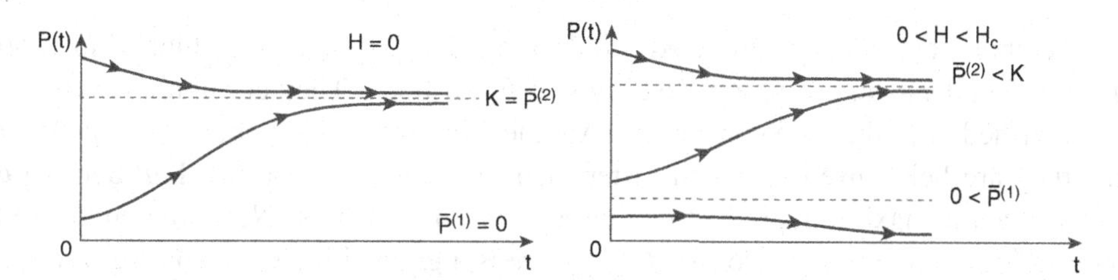

Figure 4.4 (a, b): Plots of $P(t)$ vs t for constant harvesting: (a) $H = 0$; (b) $0 < H < H_c$.

where E is independent of P and is related to the harvesting effort. Combining the first and third terms on the right-side gives the expression

$$\frac{dP}{dt} = [(a - E) - bP]\,P. \tag{4.2.30}$$

Inspection shows that two constant or equilibrium states exist; they are

$$\bar{P}^1 = 0 \text{ and } \bar{P}^2 = \frac{a - E}{b}. \tag{4.2.31}$$

Note that only for $E < a$ is there a stable second equilibrium solution; see Figure 4.4.

4.2.4 Hubbert Theory for Resources

An interesting and certainly economical important concept is the Hubbert curve. This is a "technique for estimating over time the production rate of a natural resource such as

- Coal
- Natural gas
- Helium gas
- Fish (in the ocean)
- Fissionable materials.

It is based on the assumption that all natural resources are finite or limited in quantity.

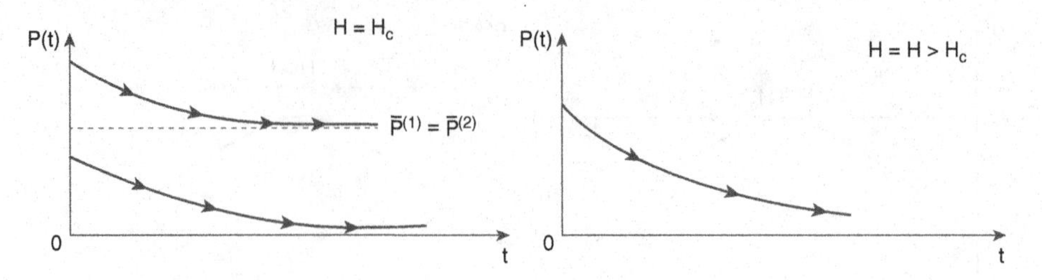

Figure 4.5 (a, b): Plots of $P(t)$ vs t for constant harvesting: (a) $H = H_c$; (b) $H > H_c$.

The Hubbert curve was formulated and developed by Marion King Hubbert in 1956 to analyze and understand the production rates of fossil fuels. They are nonrenewable, i.e., not replenished after they are consumed. (Another way of thinking about such substances is that they are being used up much faster than they can be replaced). **Hubbert's peak theory** is that the maximum production rate for a resource occurs "Near" the middle of the resource's life cycle. In other words, at first there is a gradual increase from zero resource production, followed by a rapid increase to a peak production, the Hubbert peak, which is the maximum production level. This is in turn followed by a decrease in production which becomes eventually a steep production decline because of the depletion of the resource.

The major task that Hubbert took on was to estimate the (approximate) date of the peak production rate relative to the time when the resource started to be "harvested."

To proceed further, we "derive" the so-called Hubbert equation, generally, called the Hubbert's peak equation as applied to crude oil production

Make the following definitions:

- t = time in years,
- $P(t)$ = cumulative production of oil in barrels at time t,
- $K = P(\infty)$ = ultimate amount of recoverable oil,
- $Q(t) = dP(t)/dt$ = production rate (barrels/year) at time t.

At this point Hubbert makes a critical assumption. He assumes that $Q(t)$ is determined by the following differential equation relationship

$$Q(t) \equiv \frac{dP(t)}{dt} = aP(t)\left[1 - \frac{P(t)}{K}\right]. \tag{4.2.32}$$

However, the solution to the differential equation part of Equation (4.2.32), which is the logistic differential equation, is known and given by the expression

$$P(t) = \frac{K}{1 + Ae^{-at}}, \tag{4.2.33}$$

where

$$P_0 = P(0), \quad A = \frac{K}{P_0} - 1. \tag{4.2.34}$$

Therefore, substituting Equation (4.2.33) into the right-side of Equation (4.3.32) gives

$$Q(t) = (aKA)\ \frac{e^{-at}}{(1 + Ae^{-at})^2}, \tag{4.2.35}$$

which is the Hubbert curve. Note that either from Equation (4.2.32) or (4.2.35), it follows that

$$Q(0) = aP_0\left[1 - \frac{P_0}{K}\right],\ Q(\infty) = 0. \tag{4.2.36}$$

While the Hubbert curve for estimating the peak of a resource production rate has been used, with several modifications, extensively by industry, it is clearly an ad hoe construction and has no scientific basis. The fact that it has proved useful is quite amazing in light of this fact. We stop at this point and leave it to reader to decide whether they wish to know more on this topic.

4.3 GOMPERTZ MODEL

4.3.1 Gompertz Equation and Solution

Another type of population growth model is based on the fact that Ln (1) = 0. Therefore, a mathematical model having constant solutions or equilibrium solutions at

$$\bar{P}^1 = 0,\quad \bar{P}^2 = K \tag{4.3.1}$$

is

$$\frac{dP}{dt} = -\lambda P \,\mathrm{Ln}\left(\frac{P}{K}\right), \tag{4.3.2}$$

where

$$\lambda > 0,\ K > 0,\ P(0) = P_0 > 0. \tag{4.3.3}$$

Note the appearance of a negative sign on the right-side of Equation (4.3.2). It can be eliminated if we use the following property of the logarithm function, i.e.,

$$\mathrm{Ln}\,(x) = -\mathrm{Ln}\left(\frac{1}{x}\right). \tag{4.3.4}$$

Therefore, Equation (4.3.2) can be written as

$$\frac{dP}{dt} = \lambda P \,\mathrm{Ln}\left(\frac{K}{P}\right). \tag{4.3.5}$$

This differential equation was derived by Benjamin Gompertz to model the growth of certain types of tumors.

If we denote the right-side of the Gompertz differential equation by $F(P)$, i.e.,

$$F(P) = -\lambda P \,\mathrm{Ln}\left(\frac{P}{K}\right), \tag{4.3.6}$$

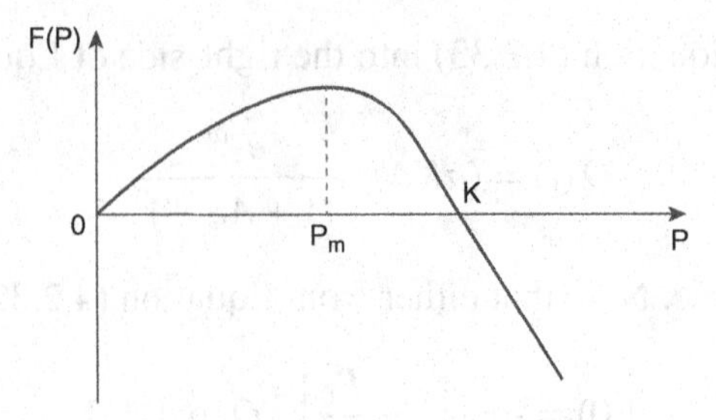

Figure 4.6: Plot of $F(P)$ vs P for the Gompertz equation. $P_m = K/e$ and $F(P_m) = \lambda$.

we can determine the location and value of its peak by taking the first-derivative and setting it to zero, i.e.;

$$\begin{aligned}\frac{dF(P)}{dP} &= -\lambda \,\mathrm{Ln}\left(\frac{P}{K}\right) - (\lambda P)\left[\frac{1}{\left(\frac{P}{K}\right)} - \frac{1}{K}\right] \\ &= -\lambda \,\mathrm{Ln}\left(\frac{P}{K}\right) - \lambda \\ &= -\lambda\left[\mathrm{Ln}\left(\frac{P}{K}\right) + 1\right],\end{aligned} \tag{4.3.7}$$

and

$$\frac{dF(P)}{dP} = 0 \Rightarrow P_m = \left(\frac{1}{e}\right)K, \tag{4.3.8}$$

where $(1/e) = 0.367879...$, with

$$F(P_m) = -\lambda \,\mathrm{Ln}\left(\frac{P_m}{K}\right) = \lambda. \tag{4.3.9}$$

See Figure 4.6 for a plot of $F(P)$ vs P.

Let us now calculate the solution to the *Gompertz* differential equation. To start, make the linear transformation

$$P = Ky, \tag{4.3.10}$$

then Equation (4.3.5) becomes

$$\frac{dy}{dt} = -\lambda y \,\mathrm{Ln}(y) \tag{4.3.11}$$

and this can be rewritten to the following form

$$\left(\frac{1}{y}\right)\frac{dy}{dt} = -\lambda \,\mathrm{Ln}(y). \tag{4.3.12}$$

But,

$$\frac{dy}{y} = d[\mathrm{Ln}(y)], \tag{4.3.13}$$

and this means that Equation (4.3.11) can be expressed as

$$\frac{d\,\mathrm{Ln}(y)}{dt} = -\lambda \,\mathrm{Ln}(y), \tag{4.3.14}$$

which has the solution

$$\text{Ln}(y) = Ae^{-\lambda t}, \tag{4.3.15}$$

where A is an arbitrary constant. Solving for the variable y gives

$$y(t) = e^{-Ae^{-\lambda t}}, \tag{4.3.16}$$

and in terms of the variable $P(t)$, we have

$$P(t) = Ke^{-Ae^{-\lambda t}}, \tag{4.3.17}$$

To determine the constant A, evaluate $P(t)$ at $t = 0$, to obtain

$$P_0 = Ke^{A} \Rightarrow A = \text{Ln}\left(\frac{P_0}{K}\right). \tag{4.3.18}$$

Note that

$$P_0 < K \Rightarrow A < 0. \tag{4.3.19}$$

Placing this A into $P(t)$ above and resulting expression finally gives the result

$$P(t) = K\left(\frac{P_0}{K}\right)^{e^{-\lambda t}}. \tag{4.3.20}$$

This is the solution to the Gompertz equation. To check its validity, it follows from Equation (4.3.20) that

$$P(0) = K\left(\frac{P_0}{K}\right) = P_0, \tag{4.3.21}$$

$$P(\infty) \lim_{t \to \infty} P(t) = K. \tag{4.3.22}$$

4.3.2 Gompertz Model with Harvesting

With harvesting the Gompertz model becomes

$$\frac{dP}{dt} = -\lambda P \,\text{Ln}\left(\frac{P}{K}\right) - EP, \tag{4.3.23}$$

where E is the harvesting effort. Factoring out P gives

$$\frac{dP}{dt} = -P\left[E + \lambda\,\text{Ln}\left(\frac{P}{K}\right)\right], \tag{4.3.24}$$

and this equation has two equilibrium solutions

$$\bar{P}^{(1)} = 0, \;\; \bar{P}^{(2)} = Ke^{-\frac{E}{\lambda}} < K. \tag{4.3.25}$$

Inspection of the expression for $\bar{P}^{(2)}$, in Equation (4.3.24), and Figure 4.6 shows that this harvesting strategy does not eliminate the tumor.

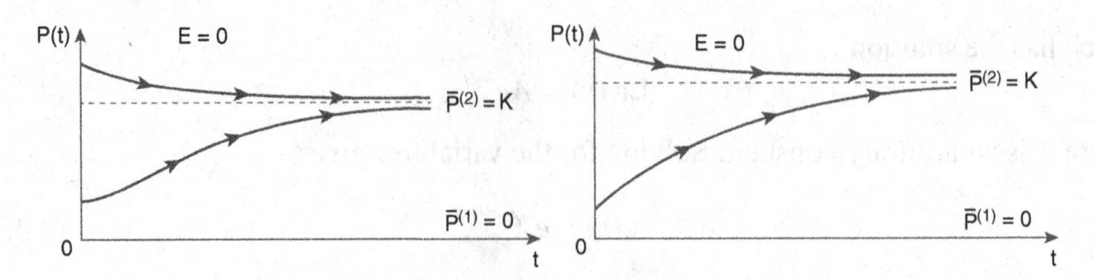

Figure 4.7: Gompertz model solutions for (a) no harvesting; (b) harvesting.

4.3.3 von Bertalanffy Equation for Tumor Growth

A rather general tumor growth model was constructed by Ludwig von Bertalanffy based on the equation having the following structure

$$\frac{dV}{dt} = aV^{\frac{2}{3}} - bV, \tag{4.3.26}$$

where $V(t)$ is the volume of the tumor at time t, and (a, b) are positive parameters. The major assumptions are that tumor growth is proportional to its surface area, while tumor material death is proportional to the volume.

Since,

$$\text{Vol} \propto L^3, \text{Area} \propto L^2, \tag{4.3.27}$$

and by assumption

$$\begin{aligned} \frac{dV}{dt} &= a_1\,(\text{Area}) - b_1\,(\text{Volume}) \\ &= a_2 L^2 - b_2 L^3, \end{aligned} \tag{4.3.28}$$

the result in Equation (4.3.26) immediately follows.

Equation (4.3.26) is Bernoulli differential equation and has the solution

$$V(t) = \left[\left(\frac{a}{b}\right) - \left(\frac{a}{b} - V_0^{\frac{1}{3}}\right) e^{-\frac{bt}{3}}\right]^3. \tag{4.3.29}$$

This solution has the same qualitative properties as the solution to the logistic equation.

4.4 NON-LOGISTIC MODELS

4.4.1 $P' = \Pi - \lambda P$

This particular model for single population growth only appears in models of coupled populations. It clears violates the "axiom of parenthood". Perhaps, its value comes from the fact that it makes the mathematical analysis easier for systems of interacting populations.

The differential equation

$$\frac{dP}{dt} = \Pi - \lambda P, \quad P(0) = P_0 > 0, \tag{4.4.1}$$

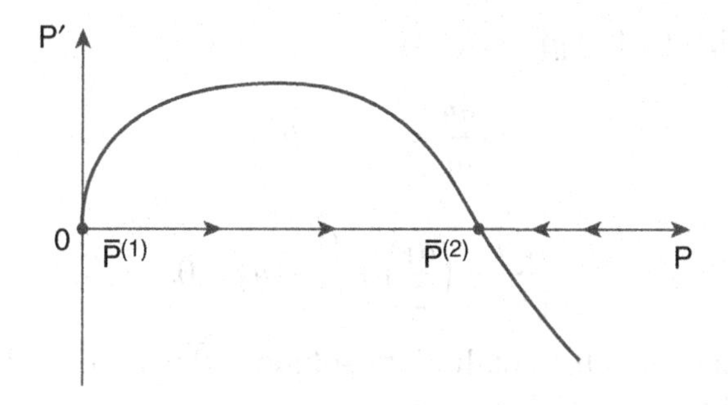

Figure 4.8: Plot of P' vs P for Equation (4.4.7).

where π is the constant net birthrate and λ is the constant deathrate, has the solution

$$P(t) = \left(P_0 - \frac{\Pi}{\lambda}\right) e^{-\lambda t} + \frac{\Pi}{\lambda}. \tag{4.4.2}$$

Note that, as expected, $P(t) = 0$ is not an equilibrium state of the system. The only such state is

$$\bar{P}^{(2)} = \frac{\Pi}{\lambda}, \tag{4.4.3}$$

and this is stable in the following sense

$$P(0) > 0 \Rightarrow P(t) > 0, \tag{4.4.4}$$

$$\lim_{t \to \infty} P(t) = \frac{\Pi}{\lambda}. \tag{4.4.5}$$

4.4.2 $P' = a_1 \sqrt{P} - b_1 P$

If we associate the first term with population births and the second with population deaths, then the average length of survival of the population members is $1/b_1$, and the birth rate for small population numbers is faster than exponential growth.

There are two equilibrium states, $\bar{P}^{(1)}$ and $\bar{P}^{(2)}$, where values are

$$\bar{P}^{(1)} = 0, \quad \bar{P}^{(2)} = \left(\frac{a_1}{b_1}\right)^2 \tag{4.4.6}$$

If we consider the (P, P') plane, then from

$$\frac{dP}{dt} = a_1 \sqrt{P} - b_1 P, \tag{4.4.7}$$

the plot in Figure 4.6 is obtained. Inspection of this plot shows that the equilibrium states $\bar{P}^{(1)}$ and $\bar{P}^{(2)}$ are, respectively unstable and stable.

The solution to this differential equation can be easily calculated if we make the following nonlinear transformation of the dependent variable

$$u(t) = \sqrt{P(t)}. \tag{4.4.8}$$

Doing this for Equation (4.4.7) gives

$$2u\frac{du}{dt} = a_1 u - b_1 u^2, \tag{4.4.9}$$

or

$$u\left[\frac{du}{dt} - \left(\frac{a_1}{2}\right) + \left(\frac{b_1}{2}\right)u\right] = 0. \tag{4.4.10}$$

Thus, $u = 0$, corresponds to the equilibrium solution, $\bar{P}^{(1)} = 0$, while the other solution come from solving the differential equation

$$\frac{du}{dt} = \left(\frac{a_1}{2}\right) - \left(\frac{b_1}{2}\right)u. \tag{4.4.11}$$

This first-order, linear equation has the solution

$$u(t) = \left[\sqrt{P_0} - \left(\frac{a_1}{b_1}\right)\right] e^{-\frac{b_1 t}{2}} + \left(\frac{a_1}{a_2}\right), \tag{4.4.12}$$

where we used the fact that $u(0) = \sqrt{P(0)}$. Therefore, since $P(t) = u(t)^2$, the solution for this single population model is

$$P(t) = \left\{\left[\sqrt{P_0} - \left(\frac{a_1}{b_1}\right)\right] e^{-\frac{b_1 t}{2}} + \left(\frac{a_1}{a_2}\right)\right\}. \tag{4.4.13}$$

4.4.3 $P' = a_2 P - b_2 P^3$

This model

$$\frac{dP}{dt} = a_2 P - b_2 P^3, \ P(0) = P_0 > 0, \tag{4.4.14}$$

has a deathrate that depends strongly on the population, i.e., it is proportional to P^3 rather than the P^2 dependence of the logistic equation.

If we multiply this equation by 2P

$$2P\frac{dP}{dt} = (2a_2)P^2 - (2b_2)P^4, \tag{4.4.15}$$

and define the new variable z to be

$$z = P^2, \tag{4.4.16}$$

then z satisfies the differential equation

$$\frac{dz}{dt} = (2a_2)z - (2b_2)z^2, \tag{4.4.17}$$

However, this is just a logistic equation whose solution is known.

Note that Equation (4.4.14) has three equilibrium solutions

$$\bar{P}^{(1)} = 0, \ \bar{P}^{(2)} = \sqrt{\frac{a_2}{b_2}}, \ \bar{P}^{(3)} = -\sqrt{\frac{a_2}{b_2}}, \tag{4.4.18}$$

where for $P_0 > 0$, the solution $\bar{P}^{(3)}$ is unphysical i.e., cannot be reached.

4.5 ALLEE EFFECT

This effect was introduced by Warder Clyde Allee to model complex behaviors that could influence population growth dynamics. For this situation to occur, the growth rate must be negative at small population. This condition can exist for a number of reasons: For example, there might be mate limitation, i.e., finding suitable mates for reproduction. Also, cooperative defense may be required and this higher state of vigilance can result in less effort spend on finding sources of food and water. Or there might be advantage in cooperative feeding and this could result in diminished food supplies when the population is small.

A logistic equation modified to show the Allee effect is

$$\frac{dP}{dt} = aP\left(1 - \frac{P}{K}\right)\left(\frac{P}{L} - 1\right), \tag{4.5.1}$$

with

$$L < K. \tag{4.5.2}$$

Figure 4.7 gives a plot of P' vs P. The arrows on the P-axis indicates the increase or decrease of the solution $P(t)$. The following information can be easily determined from this plot.

(i) There are three equilibrium states or solutions. They are located at

$$\bar{P}^{(1)} = 0, \quad \bar{P}^{(2)} = L, \quad \bar{P}^{(3)} = K, \tag{4.5.3}$$

(ii) These three equilibrium solutions and states are, respectively, (stable, unstable, stable).

(iii) The population will only sustain itself if its initial value is large enough, i.e.,

$$\begin{cases} 0 < P_0 < L : P(t) \text{ decreases to zero;} \\ L < P_0 < K : P(t) \text{ increases to } K; \\ P_0 > K \quad\quad : P(t) \text{ decreases to zero.} \end{cases} \tag{4.5.4}$$

4.6 DISCUSSION

This chapter considered the logistic equation and several of its generalizations. These differential equations provide models for the growth of a single population. Except for the representation given in Section 4.4.1, all the examined models took the form

$$\frac{dP}{dt} = f(P), \quad P(0) = P_0 > 0, \tag{4.6.1}$$

where

$$f(0) = 0, \quad f(K) = 0. \tag{4.6.2}$$

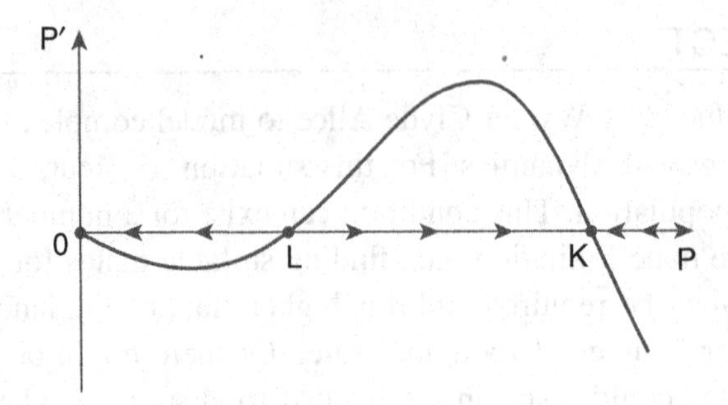

Figure 4.9: Plot of P vs P' illustrating the Allee effect. There are equilibrium state at $\bar{P}^{(1)} = 0$, $\bar{P}^{(2)} = L$, and $\bar{P}^{(3)} = K$.

The solutions

$$P(t) = \bar{P}^{(1)} = 0, \quad \bar{P}(t) = P^{(2)} = K, \tag{4.6.3}$$

are the equilibrium, constant solutions and are the steady states of the population.

The non-negative zeros of $f(P)$ correspond to equilibrium populations and mathematically two simple ways of incorporating them into $f(P)$ is to have $f(P)$ contain two factors such that one is zero at $P = 0$, while the other is zero at $P = K$. This suggests that $f(P)$ has one or other of the forms

$$f(P) = \begin{cases} g_1(P)\ P(P-K), \\ g_2(P)\ P\,\mathrm{Ln}\left(\dfrac{K}{P}\right), \end{cases} \tag{4.6.4}$$

where $g_1(P)$ and $g_2(P)$ do not have zeros for $P \geq 0$.

It should be stated that within the framework of our differential equation models, what we construct is but an approximation to the actual population dynamics. The goal of obtaining an exact mathematical model containing all features of the "real world" population dynamics is impossible. Thus, there is an inherent ambiguity in the modelling process and no possibility exists for uniqueness. The essential value of a mathematical model is to produce (predict) results that are not inputs into its construction, and to give insights on possible mechanisms that cause the system to have its particular dynamical features. Finally, it is the fact that these systems are "ill-defined" that allows for the opportunities to make future improvements in the models.

We conclude this chapter by illustrating one of the difficulties which shows up when comparing two (or more) mathematical models of single population growth.

Consider the two models

$$\text{(A)}\ \frac{dP}{dt} = aP - bP^2, \tag{4.6.5}$$

$$\text{(B)}\ \frac{dP}{dt} = a_1P^2 - b_1P^3. \tag{4.6.6}$$

By rescaling of the variables, i.e.,

$$P = P^*u, \ t = T^*\eta, \tag{4.6.7}$$

we obtain the dimensionless equations

$$(A_1)\begin{cases}\dfrac{du}{d\eta} = u - u^2,\\ P^* = \dfrac{a}{b}, \quad T^* = \dfrac{1}{a};\end{cases} \tag{4.6.8}$$

$$(B_1)\begin{cases}\dfrac{du}{d\eta} = u^2 - u^3,\\ P^* = \dfrac{a_1}{b_1}, \quad T^* = \dfrac{b_1}{a_1^2}.\end{cases} \tag{4.6.9}$$

Note that in the rescaled variables, the equilibrium solutions for both equations are at $\bar{U}^{(1)} = 0$ and $\bar{u}^{(2)} = 1$.

If the initial condition is selected to be very small, i.e.,

$$u(0) = u_0 << 1, \tag{4.6.10}$$

then dominant balance allows the use of the following two differential equations to calculate approximate solutions

$$(A_2)\,\frac{du}{d\eta} \simeq u, \quad u(0) = u_0, \tag{4.6.11}$$

$$(B_2)\,\frac{du}{d\eta} \simeq u^2, \quad u(0) = u_0. \tag{4.6.12}$$

We now ask the question: What is the approximate time for the populations to increase to the value $u(t^*) = 1/10$? Assuming that Equations (4.6.11) and (4.6.12) hold their validity, simple integrations give the estimates

$$(A_3)\ t^* \simeq \mathrm{Ln}\left(\frac{1}{10u_0}\right), (B_3)\ t^* \simeq \frac{1}{u_0}. \tag{4.6.13}$$

For $u_0 = 10^{-4}$, we find

$$(A_3)\ t_A^* \simeq 6.91, \quad (B_3)\ t_B^* \simeq 10^4. \tag{4.6.14}$$

Note that the ratio of these two dimensionless time scales is

$$\frac{t_B^*}{t_A^*} \simeq 1500. \tag{4.6.15}$$

Consequently, it would be unreasonable to plot the corresponding two solutions on the same graph using a linear time scale. However, with a logarithmic scale something like that shown in Figure 4.10 would appear.

This exercise shows that two models may have exactly the same qualitative properts for their solutions, while differing widely in the details of their quantitative values. This fact is generally true for various models of an arbitrary system.

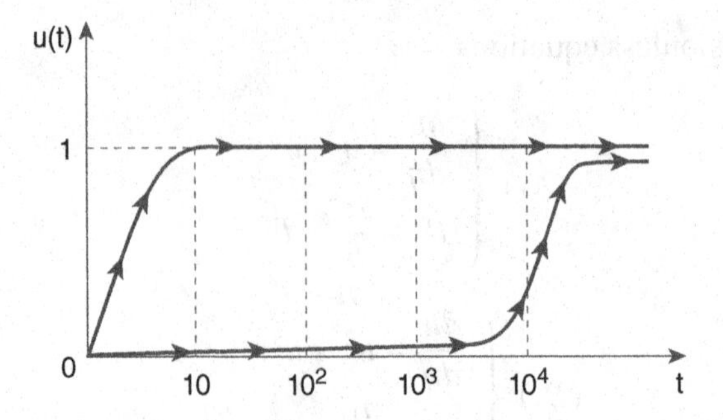

Figure 4.10: Plots of the solutions for Equations (4.6.8) and (4.6.9) for the same but very small initial conditions; $\eta = t$.

PROBLEMS

Section 4.1

1) Examine in detail the "axiom of parenthood" and its implications. What are if any, the conditions under which it might be valid to ignore it?

Section 4.2

2) Construct a mathematical form for a harvesting strategy that transitions between H and EP. Use it to study the solutions of a logistic equation which includes this term.

Section 4.3

3) Explain why the Gompertz model for tumor growth with harvesting loses its validity when $V(t)$ becomes small.

4) Assume we have a spherical shaped tumor mass. Reformulate the Gompertz differential equations in terms of the radius of the tumor rather than the volume.

   Hint: $V = \left(\frac{4}{3}\right)\pi r^3$.

Section 4.4

5) Can the model, $P' = aP^2 - bP^3$, be solved exactly?

   Hint: Look up the Lambert-W function.

NOTES AND REFERENCES

Section 4.1: An excellent broad introduction to continuous population models for a single population is

(1) J.D. Murray, **Mathematical Biology** (Springer-Verlag, Berlin, 1989). See Chapter 1.

The "axiom of parenthood" was introduced in the book

(2) G.E. Hutchinson, Introduction to Population Ecology (Yale University Press, New Haven, 1978). See Chapter 1.

The following document gives a good overview of population dynamics and a brief summary of its history

(3) A. Salisburg, **Mathematical Models in Population Dynamics** (New College of Florida; Sarasota, FL; April 2011). This is a Bachelor of Arts Thesis, submitted in partial fulfillment for this degree.

Section 4.2: The use of the logistic equation to understand the spread of innovations is discussed in

(4) M. Braun, **Differential Equations and Their Applications** (Springer-Verlag, New York, 1975). See Section 1.6.

The "derivation" of the Hubbert curve and some of its applications to actual data analysis are discussed in the next three references

(5) M. King Hubbert, Nuclear energy and fossil fuels, Publication 95, Shell Oil Company, 1956. This is located at the site http://www.hubbertpeak.com/hubbert/1956/1956.pdf

(6) A. Valero and A. Valero, Physical geonomics: Combing the energy and Hubbert peak analysis for predicting mineral depletion, **Resources, Conservation and Recycling**, Vol. 54, Issue 12 (October 2010), 1074-1083.

(7) WIKIPEDIA: "Hubbert curve" and "Hubbert peak theory."

Section 4.3: For information on the Gompertz population model see

(8) A.K. Laird, Dynamics of tumor growth, British Journal of Cancer, Vol. 3 (1964), 490

(9) C.P. Winor, The Gompertz curve as a growth curve, Proceedings of the National Academy of Science, Vol. 18 (1932), 1-8.

(10) S. Roberts, **Harvesting the Single Species Gompertz Population Model in a Slowly Varying Environment** (Australian Mathematics Science Institute, University of Melbourne; Victoria, 2014), 10 pages.

Section 4.4: This section is a summary of the publication

(11) R.E. Mickens, Mathematical and Numerical comparisons of Five single-population growth models, **Journal of Biological Dynamics**, Vol. 10, Issue 1 (2016), 95-103.

Section 4.5: A useful introduction to the Allee effect and how it may arise is

(12) Wikipedia: "Allee effect."

The original publication is

(13) W.C. Allee and E. Bowen, Studies in animal aggregations, **Journal of Experimental Zoology**, Vol. 61, Issue 2 (1932), 185-207.

See also

(14) F. Courchamp, J. Berec, and J. Gascoigne, **Allee Effects in Ecology and Conservation** (Oxford University Press, New York, 2008).

CHAPTER 5

$1 + 2 + 3 + 4 + 5 + \cdots = -(1/2)$

5.1 INTRODUCTION

In theoretical physics, there appear in a number of publications, calculations based on expressions such as

$$1 + 2 + 3 + 4 + \cdots = -\left(\frac{1}{12}\right), \tag{5.1.1}$$

and

$$1 - 1 + 1 - 1 + 1 - 1 + \cdots = \frac{1}{2}. \tag{5.1.2}$$

In the general, non-physics informal and formal research communities, a number of explanations have been given to help understand what these "relations" actual mean. The purpose of this chapter is to introduce my views on these and related issues. We present arguments in support of the idea that expressions such as those listed in Equations (5.1.1) and (5.1.2) can be understood by considering them as the result of modelling certain questions such as

"What is$1 + 2 + 3 + 4 + \cdots$?"

Based on the model selected, the answer is $(-1/12)$. In other words, we ask what does "$1 + 2 + 3 + 4 + \cdots$" mean within the context of some constructed model and then use the model to evaluate the numerical value of "$1 + 2 + 3 + 4 + \cdots$."

Similar kinds of things can be done for the relation

$$\int_0^\infty e^t dt = -1, \tag{5.1.3}$$

an example created by the author.

In outline, the next section presents brief introduction to the main ideas relating to analytic continuation and functional equations. Section 5.3 considers how convergent series should be "looked at" or evaluated when their independent variable takes values for which the series is not convergent. We give a Rule as to how this evaluation should be done.

DOI: 10.1201/9781003178972-5

Section 5.4 examines the relation given in Equation (5.1.3) and shows how this relation is obtained. The Gamma function

$$\Gamma(z) \equiv \int_0^\infty t^{z-1} e^{-t} dt, \tag{5.1.4}$$

and its major properties are presented in Section 5.5. Finally, we define and state many of the important features of the Riemann Zeta function in Section 5.6, and show why the result in Equation (5.1.1) should/is correct.

It is strongly suggested that readers who need to refreshen their knowledge of calculus and concepts related to convergent series and integrals, skim the relevant sections of the book by Mendelson and Ayres (2013). A similar comment applies to those readers who either have not taken a course in complex variables or who need to jog their memories on its fundamental principles. An excellent place to do this is the book of Spieget et al. (2009).

5.2 PRELIMINARIES

5.2.1 Functional Equations

A functional equation is an equation in which the "unknown" is a function. In general, a functional equation relates the value of a function at some particular point to its values at one or more other points.

The following are examples of functional equations:

$$f(x+y) = f(x) + f(y), \tag{5.2.1}$$

$$f(xy) = f(x) f(y), \tag{5.2.2}$$

$$f(x)^2 + g(x)^2 = 1 \tag{5.2.3}$$

$$y_{k+2} - 2y_{k+1} + y_k = 0, \; k = \text{integer}. \tag{5.2.4}$$

Possible solutions, respectively, to these equations are

$$f(x) = Ax, \tag{5.2.5}$$

$$f(x) = x^n, \tag{5.2.6}$$

$$f(x) = \cos x, \quad g(x) = \sin x, \tag{5.2.7}$$

$$y_k = B_1 + B_2\, k_3, \tag{5.2.8}$$

where (A, B_1, B_2, n) are arbitry constants. Note that the solution to a functional equation may not be unique. In particular, Equation (5.2.3) also has the solution

$$f(x) = cn(x), \quad g(x) = sn(x), \tag{5.2.9}$$

where cn/sn are the Jacobi cosine and sine functions.

An important class of functional equations of value to the concerns of this chapter are those which take the form

$$f(z+1) = G(f(z), z), \tag{5.2.10}$$

where $G(u, v)$ is a specified function of u and V. Note that knowledge of $f(z)$ over any "unit interval" in z allows $f(z)$ to be calculated for any other value of z.

Comment

If z is real, i.e., $z = x$, then the interval is a line segment on the x-axis. If z is a complex variable, the interval is a vertical strip in the z-plane.

5.2.2 Analytic Continuation

Let $z = x + iy$. A function $f(z)$ is an analytic function at z_0 if it has a convergent power series at z_0, i.e.,

$$f(z) = f(z_0) + f^{(1)}(z_0)(z - z_0) + f^{(2)}(z_0)(z - z_0)^2 + \cdots + f^{(n)}(z_0)(z - z_0)^n + \cdots, \tag{5.2.11}$$

where

$$f^{(n)}(z_0) = \left(\frac{1}{n!}\right) \left.\frac{d^n f(z)}{dz^n}\right|_{z=z_0}. \tag{5.2.12}$$

This series converges over a circular region of the complex z-plane, centred at z_0, with radius of convergence equal to the distance from z_0 to the nearest singularity of $f(z)$.

If we use $z = x + iy$, then $f(z)$ can be written as

$$\begin{aligned} f(z) &= f(x + iy) \\ &= u(x, y) + iv(x, y). \end{aligned} \tag{5.2.13}$$

The function $f(z)$ is analytic at $z_1 = x_1 + iy_1$ if u and v are continuous at (x_1, y_1) and the Cauchy- Riemann relations

$$\left.\frac{\partial u(x, y)}{\partial x}\right|_{\substack{x=x_1\\ y=y_1}} = \left.\frac{\partial v(x, y)}{\partial y}\right|_{\substack{x=x_1\\ y=y_1}}, \tag{5.2.14}$$

$$\left.\frac{\partial u(x, y)}{\partial y}\right|_{\substack{x=x_1\\ y=y_1}} = -\left.\frac{\partial v(x, y)}{\partial x}\right|_{\substack{x=x_1\\ y=y_1}} \tag{5.2.15}$$

are satisfied.

Analytic continuation is a procedure which allows the extension of the domain of an analytic function. Consider a $f(z)$ that is analytic on a region R. Let R be contained fully in a region S. The function $f(z)$ can be analytically continued from R to S if there exists a function $g(z)$ such that

(a) $g(z)$ is analytic on region S;

(b) $g(z) = f(z)$ for all z in region R.

An important feature of analytic continuation is that it is unique, i.e, if there is a second function $g_1(z)$ that satisfies conditions (a) and (b), then $g_1(z) = g(z)$.

This conclusion can be generalized to the situation where R and S overlap, and $g(z) = f(z)$ in the overlap region. See Figure 5.1 for the graphic details.

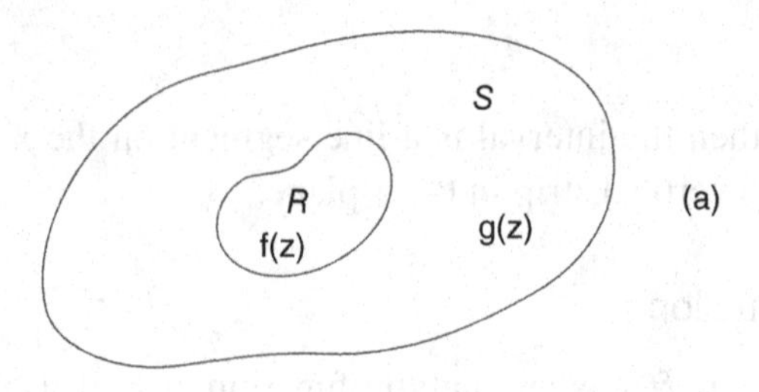

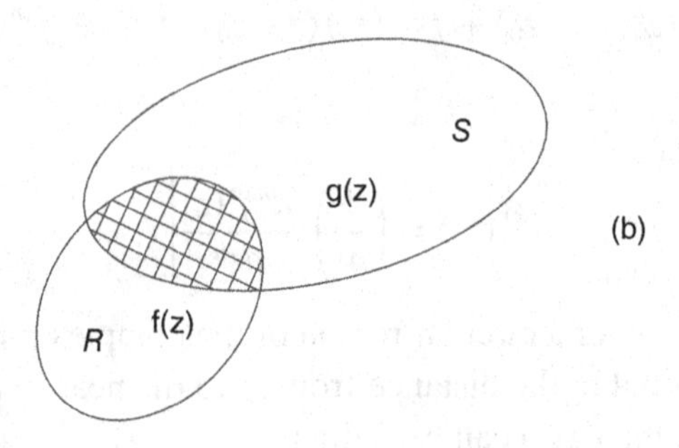

Figure 5.1: Two regions in the z-plane. (a) R is contained in S (b) R overlaps S.

5.3 NUMERICAL VALUES OF DIVERGENT SERIES

Consider the following two functions and their respective Taylor series

$$h(x) = \frac{1}{1-x} = 1 + x + x^2 + \cdots + x^n + \cdots, \tag{5.3.1}$$

$$g(x) = \frac{1}{1+x^2} = 1 - x^2 + x^4 + \cdots + (-1)^n x^{2n} + \cdots, \tag{5.3.2}$$

Both series expansions coverage for $|x| < 1$. But, what occurs for values of x such that $|x| > 1$? Clearly, the series do not converge and they are called divergent series. However, just because they are divergent series do not mean that they are meaningless or that interpretations cannot be attached to them to give **finite numerical values**. We demonstrate how this may be accomplished.

To begin, observe that

$$\frac{dh(x)}{dx} = \frac{1}{(1-x)^2}, \frac{1}{(1-x)^2} \tag{5.3.3}$$

and this has the convergent series expansion

$$\begin{cases} \dfrac{dh(x)}{dx} = 1 + 2x + 3x^2 + \cdots + nx^{n-1} + \cdots, \\ |x| < 1. \end{cases} \tag{5.3.4}$$

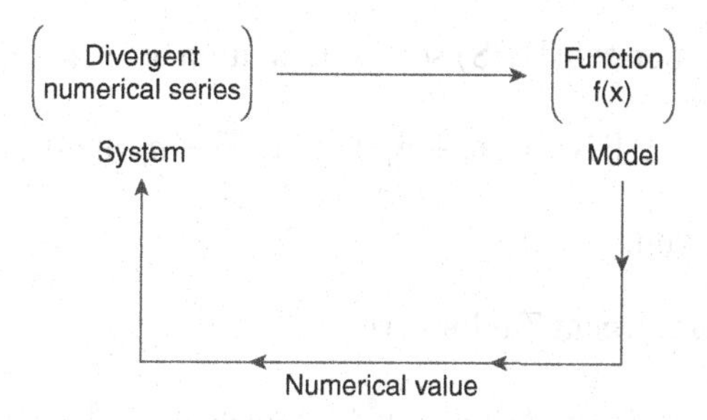

Figure 5.2: Modelling divergent numerical series as specific values of a function $f(x)$.

If the right-sides of Equations (5.3.1), (5.3.2), and (5.3.4) are evaluated at either $x = -1$ or $x = 1$, then the following results are obtained,

$$h(-1) = \frac{1}{2} = 1 - 1 + 1 - 1 + 1 - 1 + \cdots, \tag{5.3.5}$$

$$g(+1) = g(-1) = \frac{1}{2} = 1 - 1 + 1 - 1 + 1 - 1 + \cdots, \tag{5.3.6}$$

$$h'(-1) = \frac{1}{4} = 1 - 2 + 3 - 4 + \cdots + (-1)^{n+1} n. + \cdots. \tag{5.3.7}$$

Note that we have introduced a new symbol

$$=, \tag{5.3.8}$$

meaning the divergent series is assigned a numerical value given by the specified functions $g(x)$ and $h(x)$ at the particular values (± 1).

Within the context of this point of view, the assignment of a finite numerical value to a divergent series is based on the ability to find a function, which when evaluated at a specific value of its variable, reproduces the divergent numerical series.

In effect, returning to the results in Equation (5.3.1), we are assuming that the series on the right-side can be assigned a numerical value for any x for which $h(x)$ is defined. For example, for $x = 2$, we have

$$h(2) = \frac{1}{1-2} = -1, \tag{5.3.9}$$

and consequently,

$$1 + 2 + 4 + 8 + \cdots + 2^n + \cdots = (-1). \tag{5.3.10}$$

Figure 5.2 presents this procedure as a modelling process:

(A) Select a divergent numerical series to "evaluate."

(B) Find a function, $f(x)$, such that for $x = x^*$, its extended Taylor series gives the divergent numerical series.

(C) Assign to the divergent numerical series the value $f(x^*)$.

As a modelling process, the system is the original divergent numerical series and the function, $f(x)$, is a mathematical model of it, for $x = x^*$.

In somewhat more details, the above procedure goes as follows:

(a) Let the divergent numerical (DNS) series be written as

$$\text{DNS} = a_0 + a_1 + a_2 + \cdots + a_n + \cdots, \tag{5.3.11}$$

where the $\{a_i\}$ are given,

(b) Let $f(x)$ have the convergent Taylor series

$$f(x) = f_0 + f_1 x + f_2 x^2 + \cdots + f_n x^n + \cdots, \tag{5.3.12}$$

where the function $f(x)$, on the left-side, is known in an explicit form as a finite combination of (in general) the elementary functions.

(c) Let, for $x = x^*$, the following conditions hold

$$f_n (x^*)^n = a_n, \quad n = (0, 1, 2, \ldots). \tag{5.3.13}$$

(d) The numerical value of the DNS is taken to be or assigned the value

$$\text{DNS} = f(x^*). \tag{5.3.14}$$

It is important to understand and make sure that $x = x^*$ is a point where $f(x^*)$ exists.

To end this section, we present the following bit of lore from the toolkit of many theoretical and or mathematical physical scientists:

Mathematicians assume that a series diverges, until **proven** otherwise.
A **physical scientist** assumes that a series converges, until proven otherwise and then **reinterprets** the series to make physical sense.

It is not certain (to me) who first stated this perspective on the differing attitudes of mathematicians and theoretical physical scientists with regard to series in particular and to rigorous proof in general. But it is astonishing how this methodology has worked in the construction of valid models for the physical sciences. Mimicking Eugene Wigner, we marvel at the unreasonable effectiveness of these types of procedures when applied to the physical sciences. Particular important examples include quantum electrodynamics and string theory.

5.4 ELEMENTARY FUNCTION DEFINED BY AN INTEGRAL

5.4.1 The Function $I(z)$

Consider a function $I(z)$ defined as follows

$$I(z) \equiv \int_0^\infty e^{-zt} dt, \quad z = \text{fixed}. \tag{5.4.1}$$

If we take $z = x + iy$, then

$$|e^{-zt}| = e^{-xt}|e^{iyt}| = e^{-xt} \tag{5.4.2}$$

Therefore, the integral is convergent for

$$x = \text{Re } z > 0. \tag{5.4.3}$$

Making the transformation

$$u = zt \tag{5.4.4}$$

gives

$$I(z) = \left[\int_0^\infty e^{-u}du\right]\left(\frac{1}{z}\right), \tag{5.4.5}$$

and using

$$\int_0^\infty e^{-u}du = 1, \tag{5.4.6}$$

the result

$$I(z) = \frac{1}{z}, \quad \text{Re } z > 0. \tag{5.4.7}$$

Let $g(z)$ be the complex function

$$G(z) = \frac{1}{z}, \quad z \in [C - \{0\}], \tag{5.4.8}$$

where the notation in the square bracket means that $G(z)$ is defined in the whole complex z-plane, except at the origin, $z = 0$. The regions of existence for $I(z)$ and $G(z)$ are shown in Figure 5.3, where it is seen that they overlap for Re $z > 0$. Thus, using the results presented in Section 5.2, it can be concluded that $G(z)$ is the analytic continuation of $I(z)$ from Re $z > 0$ to the whole z-plane, minus the origin.

With this result, we now associate the integral, given in Equation (5.4.1) with the value $1/z$, i.e.,

$$\int_0^\infty e^{-zt}dt = \frac{1}{z}, \tag{5.4.9}$$

for all z except $z = 0$. One consequence of this association is that for $z = -1$, we have

$$\int_0^\infty e^{t}dt = -1. \tag{5.4.10}$$

Note that the left-side is not a convergent integral. Further, if we consider this integral as representing the area under the cure $f(t) = \exp(-t)$, then the usual understanding is that the area is positive and unbounded. However, our interpretation differs in that we obtain a bounded, negative value. This outcome is similar to what was found for several non-convergent series discussed in Section 5.3.

These "calculations" demonstrate that just because an integral is not convergent does not imply that it is meaningless. It can have a well defined meaning through the use of another interpretation.

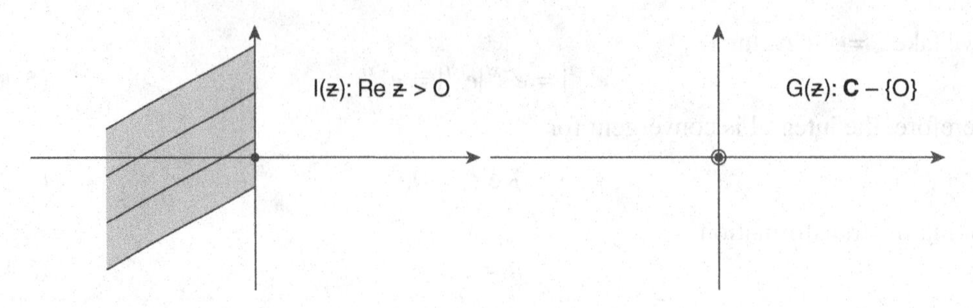

Figure 5.3: Regions of definition for $I(z)$ and $G(z)$. (a) $I(z)$ exists for Re $z > 0$. (b) $G(z)$ exists in the entire z-plane, minus the point $z = 0$.

5.4.2 Functional Relationship for $I(z)$

For Re $z > 1$, we know that the convergent integral, Equation (5.4.1), is equal to $1/z$, i.e.,

$$I(z) \equiv \int_0^\infty e^{-zt} dt = \frac{1}{z}, \quad \text{Re } z > 0. \tag{5.4.11}$$

A functional equation will now be derived for $I(z)$. First, note that from

$$I(z) = \frac{1}{z} \Rightarrow I(z+1) = \frac{1}{z+1}, \tag{5.4.12}$$

it follows that

$$I(z+1) = \frac{\left(\frac{1}{z}\right)}{1 + \left(\frac{1}{z}\right)}, \tag{5.4.13}$$

and

$$I(z+1) = \frac{I(z)}{1 + I(z)}, \quad \text{Re } z > 0, \tag{5.4.14}$$

we select the following condition for $I(z)$ to satisfy

$$I(1) = 1. \tag{5.4.15}$$

From Equation (5.4.14), we immediately observe that if $I(z)$ is known in any vertical strip in the z-plane,

$$z_0 \le z \le z_0 + 1, \tag{5.4.16}$$

then, it can be determined in any other vertical strip,

$$z_0 + N \le z \le z_0 + 1 + N, \tag{5.4.17}$$

where N is any other (positive or negative) integer. Consequently, the functional equation can be used to calculate $I(z)$ for all z. In particular, let $z = 0$, then

$$I(0+1) = \frac{I(0)}{1 + I(0)}. \tag{5.4.18}$$

But $I(1) = 1$ and therefore

$$1 = \frac{I(0)}{1+I(0)}, \tag{5.4.19}$$

which can only hold if $I(0)$ is unbounded. Therefore, $I(z)$ has a (pole) singularity at $z = 0$.

Finally, an elementary calculation shows that

$$I(z+u) = \frac{I(z)}{1+uI(z)}. \tag{5.4.20}$$

5.5 GAMMA FUNCTION

5.5.1 Derivation and Properties of the Gamma Function

The gamma function, $\Gamma(z)$, appears in many areas of applied and pure mathematics, and in the mathematical sciences. Here, we present a brief derivation of $\Gamma(z)$ list some of its important
properties.

To begin, define $f(a)$ to be

$$f(a) \equiv \int_0^\infty e^{-at} dt, \quad \text{Re}(a) > 0. \tag{5.5.1}$$

Since the value of the convergent integral is $1/a$, we have

$$\frac{1}{a} = \int_0^\infty e^{-at} dt. \tag{5.5.2}$$

Differentiating both sides of this expression, with respect to a, gives

$$\frac{(-1)^n n!}{a^{n+1}} = (-1)^n \int_0^\infty t^n e^{-at} dt. \tag{5.5.3}$$

For $a = 1$, we have

$$n! = \int_0^\infty t^n e^{-t} dt, \quad n = (0, 1, 2, 3 \cdots). \tag{5.5.4}$$

Note that

$$n! = n(n-1)\cdots 2.1, \tag{5.5.5}$$

$$n! = n(n-1)! \tag{5.5.6}$$

$$0! = 1, \quad 1! = 1. \tag{5.5.7}$$

The gamma function for z complex is defined as

$$\Gamma(z) \equiv \int_0^\infty t^{z-1} e^{-t} dt, \quad \text{Re}\, z > 0. \tag{5.5.8}$$

If z is a positive integer, $z = n$, then

$$\Gamma(n) = (n-1)! \tag{5.5.9}$$

and it follows, see Equation (5.5.6), that

$$\Gamma(n+1) = n\,\Gamma(n)\,. \tag{5.5.10}$$

This relation also holds for $\Gamma(z)$. This is a consequence of integrating $\Gamma(z+1)$ using integration by parts, i.e.,

$$\begin{aligned}\Gamma(z+1) &= \int_0^\infty t^z e^{-t} dt \qquad (5.5.11)\\ &= \left[-e^{-t} t^z\right]\Big|_{t=0}^{t=\infty} + z \int_0^\infty t^{z-1} e^{-t} dt,\end{aligned}$$

and, finally,

$$\Gamma(z+1) = z\,\Gamma(z)\,,\ \mathrm{Re}(z) > 0. \tag{5.5.12}$$

Note that this relation, a functional equation, was obtained for Re $z > 0$, we may take it to hold for all values of z for which $\Gamma(z)$ exists. Since, for n a positive integer.

$$\Gamma(z+n) = (z+n-1)(z+n-2)\cdots(z+1)\,z\,\Gamma(z)\,, \tag{5.5.13}$$

It follows that

$$\Gamma(z) = \frac{\Gamma(z+n)}{z(z+1)\cdots(z+n-1)}, \tag{5.5.14}$$

and we can use this expression to extend $\Gamma(z)$ to $\mathrm{Re}(z) > (-n)$, where n is an arbitrary positive integer.

Two important features flow, respectively, from Equation (5.5.4) and (5.5.14):

(i) $\Gamma(n) > 0$, for $\mathrm{Re}(z) = n \geq 1$.

(ii) $\Gamma(z)$ has simple zeros at $z = 0$ and the negative integers.

Also, the following product formula holds for the gamma function

$$\Gamma(z)\ \Gamma(-z) = \frac{(-1)\pi}{z \sin(\pi z)}. \tag{5.5.15}$$

Further, if the gamma function is known for any vertical strip in the complex z-plane of width one in $\mathrm{Re}(z) = x$, the Equation (5.5.12) can be used to calculate or determine $\Gamma(z)$ for any other unit vertical strip.

5.5.2 A Class of Important Integrals

A large number of the definite integrals appearing in the mathematical physical sciences have the form

$$I(\alpha, \beta, \gamma) = \int_0^\infty t^\alpha \exp(-\beta\, t^\gamma)\, dt, \tag{5.5.16}$$

where for convergence the following restrictions hold

$$\text{Re}(\alpha) > -1, \quad \text{Re}\,\beta > 0, \quad \gamma \text{ real and positive.} \tag{5.5.17}$$

This integral can be expressed in terms of the gamma function. To show this, make the variable change

$$x = \beta\, t^\gamma \text{ or } t = \left(\frac{x}{\beta}\right)^{\frac{1}{\gamma}}. \tag{5.5.18}$$

Therefore,

$$x = 0 \;\Rightarrow t = 0, \quad x = \infty \Rightarrow t = \infty, \tag{5.5.19}$$

and

$$t^\alpha = \left(\frac{x}{\beta}\right)^{\frac{\alpha}{\gamma}}, \; dt = \left(\frac{1}{\beta^{\frac{1}{\gamma}}}\right)\left(\frac{1}{\gamma}\right) x^{\frac{1-\gamma}{\gamma}}\, dx. \tag{5.5.20}$$

If these results are substituted into the right-side of Equation (5.5.16) and we find after simplification the expression

$$I = (\alpha, \beta, \gamma) = \left(\frac{1}{\gamma\beta^{\frac{\alpha+1}{\gamma}}}\right) \int_0^\infty x^{\frac{\alpha+1}{\gamma}-1} e^{-x} dx. \tag{5.5.21}$$

The integral is a gamma function, with (5.5.22)

$$z = \frac{\alpha + 1}{\gamma}, \tag{5.5.22}$$

and we finally obtain

$$I = (\alpha, \beta, \gamma) = \left[\frac{1}{\gamma\beta^{\frac{\alpha+1}{\gamma}}}\right] \Gamma\left(\frac{\alpha + 1}{\gamma}\right). \tag{5.5.23}$$

Note that using Equation (5.5.23), the original integral, Equation (5.5.16), can be analytically extended into values of the parameters, $(\alpha,\ \beta,\ \gamma)$, for which it is not convergent within the framework of standard mathematics.

5.6 RIEMANN ZETA FUNCTION

5.6.1 Definition of Riemann zeta Function

The Riemann zeta function is defined by the following infinite series

$$\zeta(z) \equiv \sum_{n=1}^{\infty} \frac{1}{n^z} = 1 + \frac{1}{2^z} + \frac{1}{3^z} + \cdots. \tag{5.6.1}$$

Using standard tests for convergence, this function converges for $\text{Re}(z) > 1$.

The following is a list of some of the fundamental properties of $\zeta(z)$:

(a) For z real $z = x$, $\zeta(x)$ is positive,

$$\zeta(x) > 0, \quad x > 1. \tag{5.6.2}$$

(b) $\zeta(z)$ for real z, satisfies the bounds,

$$\zeta(x+1) < \zeta(x) < \zeta(x-1), x > 2, \tag{5.6.3}$$

$$1 < \zeta(x) < 2, \quad x > 2. \tag{5.6.4}$$

(c) For large x, $\zeta(z)$ approaches 1, i.e.,

$$\zeta(x) = 1 + O\left(\frac{1}{2^x}\right). \tag{5.6.5}$$

(d) $\zeta(x)$ is a monotonic decreasing function for $x > 1$, i.e.,

$$\zeta(x_1) > \zeta(x_2), x_2 > x_1 \tag{5.6.6}$$

(e) The following limit holds

$$\operatorname*{Lim}_{x \to 1^+} \zeta(x) = \infty. \tag{5.6.7}$$

(f) Let P be a prime number, i.e.,

$$P = \{2, 3, 5, 7, 11, 13, 17, 19, \ldots\}, \tag{5.6.8}$$

then the Riemann zeta function has the following product representation

$$\zeta(z) = \left[\prod_{P=2}^{\infty} \left(1 - \frac{1}{P^z}\right)\right]^{-1} \tag{5.6.9}$$

(g) $\zeta(z)$ is defined for all complex values of z by means of the analytic continuation of this function for $\text{Re}(z) > 1$. This analytic continuation is unique and excludes the point $z = 1$, which corresponds to a simple pole, i.e.,

$$\operatorname*{Lim}_{z \to 1} (z-1)\zeta(z) = 1. \tag{5.6.10}$$

(h) $\zeta(z)$ satisfies a reflection functional equation given by the following formula

$$\zeta(z) = \left(2^z \pi^{z-1}\right) \sin\left(\frac{\pi z}{2}\right) \Gamma(1-z) \zeta(1-z). \tag{5.6.11}$$

(i) Particular special values of $\zeta(z)$ are

$$\zeta(0) = \frac{1}{2}; \ \zeta(-2n) = O, \ \text{n} = 1, 2, 3, \cdots; \tag{5.6.12}$$

$$\begin{cases} \zeta(-1) = -\dfrac{1}{12}, \\ \zeta(-3) = -\dfrac{1}{120}, \\ \quad \vdots \qquad \quad \vdots \\ \zeta(-13) = -\dfrac{1}{12}, \end{cases} \tag{5.6.13}$$

5.6.2 Applications

Using the series definition of the Riemann zeta function and the specific values given in Equation (5.6.13), we can evaluate several infinite·numerical series. Doing so gives the following results:

$$\zeta(0) \equiv 1 + 1 + 1 + 1 + \cdots = \left(\frac{1}{2}\right), \tag{5.6.14}$$

$$\zeta(-1) \equiv 1 + 2 + 3 + 4 + \cdots = -\left(\frac{1}{12}\right), \tag{5.6.15}$$

$$\zeta(-2) \equiv 1 + 2^2 + 3^2 + 4^2 + \cdots = O. \tag{5.6.16}$$

5.7 DISCUSSION

Mathematics is a very broad discipline and involves much more than just proving theorems. Moreover, its concepts evolve and become broader in scope, more complex, and more interesting. In particular, new interpretations arise. Within in context, this chapter introduced some techniques for looking at infinite series which converge over a finite interval of values for their independent variables. Our concern is in values of these variables for which the series do not converge. Such series are called non-convergent or divergent. Thus, the main task was to demonstrate that these series may have finite numerical values for specific numerical values of the independent variables. The general interest in such divergent series come from the realization that they arise in many of the calculations occurring in the mathematical theoretical structures of the physical sciences.

A hint as to how to proceed with these issue is to consider the possibility that a divergent series should be taken as a **single entity** and not to be obtained as a succession of partial sums, i.e.,

$$\text{“}1 + 2 + 3 + 4 + \cdots\text{”} \neq \lim_{N \to \infty} S_N, \tag{5.7.1}$$

where

$$S_N = 1 + 2 + 3 + \cdots + N. \tag{5.7.2}$$

However, one thing such be clear: There are no actual infinities appearing in science (physics). Consequently, when such artifacts do occur in a physical calculation or theory, there must exist a methodology which allows for an interpretation of them as having finite values.

PROBLEMS

Section 5.1

1) Discuss why (in mathematics) the concept of convergence is important, Include both series and definite intergrals.

2) What is a counter-intuitive situation? What roles does this concept play in mathematics and the physical sciences?

Section 5.2

3) Solve the functional equation

$$E(z+1) = \frac{E(z)}{1+E(z)}, \quad E(1) = 1$$

Does it have a unique solution?

4) Construct an "approximate differential equation" corresponding to the functional equation of Problem 3.

5) Determine some of the periodic solutions to the equations

$$f(t)^2 + g(t)^2 = 1.$$

Section 5.4

6) Standard tables of definite integrals give the two formulas

$$\int_0^\infty e^{-ax} \cos(bx)dx = \frac{a}{a^2+b^2},$$

$$\int_0^\infty e^{-ax} \sin(bx)dx = \frac{b}{a^2+b^2}.$$

Both integrals converge for Re $(a) > 0$ and b real. Discuss the case where $a = -1$. Note that for $b = O$, the first relationship reduces to

$$\int_0^\infty e^{-ax} dx = \frac{1}{a}.$$

Section 5.5

7) Show that the follow relation is true

$$B(P,q) \equiv \int_0^1 t^{P-1}(1-t)^{q-1}$$

$$= \frac{\Gamma(P)\Gamma(q)}{\Gamma(P+q)}.$$

What restrictions should be placed on p and q? $B(P,q)$ is the beta function.

Section 5.6

8) Prove the following relations for the Riemann zeta functions

$$\bullet\ \zeta(z;\ -) \equiv \sum_{k=1}^{\infty} \frac{(-1)^{k+1}}{k^z}$$
$$= \zeta(z)\left[1 - \frac{1}{2^{z-1}}\right]$$

$$\bullet\ \zeta(z;\ \text{odd}) \equiv \sum_{k=1}^{\infty} \frac{1}{(2k-1)^z}$$
$$= \zeta(z)\left[1 - \frac{1}{2^z}\right].$$

NOTES AND REFERENCES

Section 5.1: The role of divergent series in field theory is discussed in the two books

(1) A. Zee, **Quantum Field Theory in a Nutshell** (Princeton University Press, Princeton, 2003). See pps. 65-67.

(2) J. Polchinski, String Theory, Vol. 1 (Cambridge University Press, New York, 1998). See pps. 22.

Section 5.2: The following three books provide readable presentation, respectively, of topics on convergence of series, complex variables, and functional equations.

(3) E. Mendelson and F.J Ayres, **Calculus, 6th Edition** (McGraw-Hill, New York, 2013).

(4) M.R. Spiegel, S. Lipschutz, J.J S chiller, and D. Spellman, **Schaum's Outline of Complex Variables** (McGraw Hill, New York, 2009).

(5) A. Janos, **Lectures on Functional Equations and Their Applications** (Academic Press, New York, 1966).

Section 5.3: A quick introduction to divergent series is the following Wikipedia entry

(6) Wikipedia: "Divergent Series."

For a more detailed and advanced discussion see Balser's book

(7) W. Balser, **From Divergent Power Series to Analytic Functions** (Springer-Verlag, Berlin,1994).

Sections 5.5 and 5.6: Good, quick introductions to the Gamma and Riemann zeta functions, their major properties, and some of their applications appear in the book

(8) R.E. Mickens, **Mathematical Methods for the Natural and Engineering Sciences** (World Scientific, London, 2004). See Chapter 3.

Further, more detailed comments and proofs relating to the Riemann zeta function are given in the book

(9) H.M. Edwards, **Riemann's Zeta Function** (Dover; Mineota, NY; 1974).

CHAPTER 6

A Truly Nonlinear Oscillator

6.1 INTRODUCTION

Nonlinear oscillatory systems are ubiquitous in the natural universe. For many such systems the equations of motion take the mathematical form

$$m\frac{d^2x}{dt^2} + kx = \in \mathrm{F}\left(x, \frac{dx}{dt}\right), \tag{6.1.1}$$

where m is the effective mass of the system, k represents an effective "linear" spring constant, $F(\ldots,\ldots)$ is a nonlinear force function, and $\in$ is a small dimensionless parameter. Two important examples are

$$F_1 = -\lambda\, x^3, \tag{6.1.2}$$

corresponding to the Duffing equation, and

$$F_2 = \left(1 - x^2\right)\frac{dx}{dt}, \tag{6.1.3}$$

which models the van der Pol oscillator.

Equation (6.1.1) is the equation of motion for the classical or standard nonlinear oscillators. However, in general, this differential equation can not be solved exactly and expressed as a finite combination of the elementary functions. Approximations to the solutions are often based on the assumption that they can be expressed as expansions in the parameter $\in$, where

$$0 < \in \ll 1, \tag{6.1.4}$$

and

$$x(\in, t) = x_0(t) + \in x_1(t) + \in^2 x_2(t) + \cdots, \tag{6.1.5}$$

with the initial condiations

$$x(\in, 0) = x_0(0) + \in x_1(0) + \in^2 x_2(0) + \cdots = A, \tag{6.1.6}$$

$$\frac{dx(\in, o)}{dt} = \frac{dx_0(0)}{dt} + \in \frac{dx_1(0)}{dt} + \in^2 \frac{dx_2(0)}{dt^2} + \cdots = 0. \tag{6.1.7}$$

DOI: 10.1201/9781003178972-6

In most cases, the physically useful solutions to Equation (6.1.1) are bounded and either periodic or oscillatory. However, unless one is careful, the ansatz, given by Equation (6.1.5), gives solutions which are not bounded, i.e., the $x_i(t)$, for $i = (1, 2, 3, \ldots)$, increase with t. Such terms are called **secular terms** and a major task of expansion in ϵ techniques is to determine how $x(\epsilon, t)$ can be rewritten such that secular terms do not appear.

Note that all of the expansion in ϵ or perturbation methods are based on the fact that for $\epsilon = 0$ the equation of motion reduces to that for the simple harmonic oscillator

$$\epsilon = 0: \quad m\frac{d^2x}{dt^2} + kx = 0. \tag{6.1.8}$$

In recent years, attention has turned to the examination of nonlinear oscillators which do not have a harmonic oscillator limit. These oscillator may be written as

$$m\frac{d^2x}{dt^2} + \lambda\,[\text{sign}\,(x)]\,|x|^p = \epsilon\, F\left(x, \frac{dx}{dt}\right), \tag{6.1.9}$$

where $(m, \lambda, p, \epsilon)$ are positive parameters, with

$$0 < p < 1. \tag{6.1.10}$$

The first of these non-classical oscillators was

$$m\frac{d^2x}{dt^2} + \lambda x^{\frac{1}{3}} = 0. \tag{6.1.11}$$

Note that for the $p = \frac{1}{3}$ case, we have $x^{\frac{1}{3}}$ and this function is odd; therefore, no absolute value and sign (x) symbols are needed.

Inspection of Equation (6.1.9) indicates that it has no linear harmonic oscillator limit when $\epsilon \to 0$. This means that none of the standard expansion in ϵ techniques can be used to determine approximations to its solutions.

The major goal of this chapter is to determine approximations to the solutions of Equation (6.1.11), using two non-preparative techniques: the method of harmonic balance and a iterative procedure.

We demonstrate in the next section how to calculate the exact period for the nonlinear oscillator given by Equation (6.1.11) after showing that all its solutions are periodic. In Section 6.3, we determine approximate solutions using two different non-perturbative techniques and compare their accuracy. We end the chapter with a brief summary. The problems provide a venue for extending this work.

6.2 GENERAL PROPERTIES OF EXACT SOLUTIONS

6.2.1 Preliminaries

Let us rescale Equation (6.1.11) such that the new variables are dimensionless. To do so, write this equation as

$$m\frac{d^2u}{dt^2} + \lambda u^{\frac{1}{3}} = 0. \tag{6.2.1}$$

with initial conditions

$$u(0) = u_0 > 0, \quad \frac{du(0)}{dt} = 0. \tag{6.2.2}$$

Now introduce the variables x and $\bar{t}$ by means of the linear transformations

$$u = U\,x, \quad t = T\,\bar{t}, \tag{6.2.3}$$

where U and T are the u and t scales. If these are substituted into Equation (6.2.1), then we have

$$\left(\frac{mU}{T^2}\right)\frac{d^2x}{d\bar{t}^2} + \left(\lambda\, U^{\frac{1}{3}}\right)x^{\frac{1}{3}} = 0, \tag{6.2.4}$$

and

$$\frac{d^2x}{dt^2} + \left(\frac{\lambda T^2}{mU^{\frac{2}{3}}}\right)x^{\frac{1}{3}} = 0. \tag{6.2.5}$$

If we select the coefficient of $x^{\frac{1}{3}}$ to be one, then the time scale is

$$T = \left(\frac{mU^{\frac{2}{3}}}{\lambda}\right)^{\frac{1}{2}}. \tag{6.2.6}$$

The natural scale for u is

$$U = u_0, \tag{6.2.7}$$

which gives, see the first of Equation (6.2.3),

$$x(0) = 1. \tag{6.2.8}$$

Finally, dropping the bar over $\bar{t}$, we have

$$\frac{d^2x}{dt^2} + x^{\frac{1}{3}} = 0; \quad x(0) = 1, \frac{dx(0)}{dt} = 0. \tag{6.2.9}$$

Observe that in the u-variable, the time scale, T, is proportional to $(u_0)^{\frac{1}{3}}$, i.e.,

$$T = \left(\frac{m}{\lambda}\right)^{\frac{1}{2}}(u_0)^{\frac{1}{3}}. \tag{6.2.10}$$

6.2.2 Phase Space Analysis

Equation (6.2.9) can be written as a system of two first-order differential equation, i.e.,

$$\frac{dx}{dt} = y, \quad \frac{dy}{dt} = -x^{\frac{1}{3}}, \tag{6.2.11}$$

where the first expression defines the variable y. Therefore, in the z-dim (x, y) phase space, the trajectories, $y = y(x)$, are solutions to the following first-order differential equation

$$\frac{dy}{dx} = -\frac{x^{\frac{1}{3}}}{y}. \tag{6.2.12}$$

This results is a consequence of

$$\frac{dy(x)}{dt} = \frac{dy}{dx}\frac{dx}{dt}, \tag{6.2.13}$$

and

$$\frac{dy}{dx} = \frac{dy/dt}{dx/dt} = \frac{(-)x^{\frac{1}{3}}}{y}. \tag{6.2.14}$$

The equilibrium state, where $dx/dt = 0$ and $dy/dt = 0$, is

$$(x^*,\ y^*) = (0, 0)\,. \tag{6.2.15}$$

With the initial conditions

$$x(0) = 0, \quad \frac{dx(0)}{dt} = y(0) = 0, \tag{6.2.16}$$

Equation (6.2.12) can be integrated to give

$$\frac{y^2}{2} + \left(\frac{3}{4}\right)x^{\frac{4}{3}} = \frac{3}{4}. \tag{6.2.17}$$

This result follows from first rewriting Equation (6.2.12) as

$$ydy = -x^{\frac{1}{3}}\,dx, \tag{6.2.18}$$

integrating to obtain

$$\int ydy = \frac{y^2}{2} + C_1, \quad \int x^{\frac{1}{3}}dx = \left(\frac{3}{4}\right)x^{\frac{4}{3}} + C_2, \tag{6.2.19}$$

where C_1 and C_2 are integration constants, and combining these two expressions to get

$$\frac{y^2}{2} + \left(\frac{3}{4}\right)\,x^{\frac{4}{3}} = C = C_1 + C_2. \tag{6.2.20}$$

For $t = 0$, $y(0) = 0$ and $x(0) = 1$, and therefore $C = 3/4$, and Equation (6.2.17) is the outcome. This equation is often called a first-integral of Equation (6.2.9) and corresponds to physically to the total mechanical energy of an oscillator modelled by Equation (6.2.9).

Note that Equation (6.2.17) is invariant with respect to the following three transformations in the x-y phase plane:

$$\begin{cases} T_1 : x \to -x,\ y \to y, \\ T_2 : x \to x,\ y \to -y, \\ T_3 : x \to -x,\ y \to -y, \end{cases} \tag{6.2.21}$$

where

(a) T_1 corresponds to reflection in the y-axis;

(b) T_2 corresponds to reflection in the x-axis;

(c) $T_3 = T_1\, T_2 = T_2\, T_1$ corresponds to reflection through the origin.

Finally, the fact that we have an explicit expression for the trajectories in phase space, from the first integral, this allows us to conclude that all the trajectories are closed curves. Since a closed curve in phase space corresponds to a periodic solution, we further conclude that all solutions to Equation (6.2.9) are periodic. An elementary geometric proof of this is given in Mickens (2010). In particular, see Sections 2.1 and 2.2.

6.2.3 Calculation of Exact Period

We now show that the nonlinear differential equation modelling a nonlinear oscillating system will allow us to calculate its exact period. To do this, we select a more general set of initial conditions, i.e., $x(0) = 1$ is replaced by $x(0) = A$. Therefore, Equation (6.3.9) becomes

$$\frac{d^2x}{dt^2} + x^{\frac{1}{3}} = 0; \quad x(0) = A > 0, \quad \frac{dx(0)}{dt} = 0, \tag{6.2.22}$$

where A is an arbitrary positive value for the initial position, $x(0)$. The corresponding first-integral or energy function is

$$\frac{y^2}{2} + \left(\frac{3}{4}\right)x^{\frac{4}{3}} = \left(\frac{3}{4}\right)A^{\frac{4}{3}}. \tag{6.2.23}$$

A representation of this curve is given in Figure 6.1.

If the system starts at $t = 0$ at the phase space point, $(A, 0)$, then it moves along the indicated path clockwise. Based on the symmetry transformations, given in Equation (6.2.21), the time to go from $x = A$ to $x = 0$, is $(1/4)$ the time to return back to $x = A$. This total time is the period of the oscillatory motion and we denote it by T.

From Equation (6.2.23), we obtain

$$y^2 = \left(\frac{3}{2}\right)\left(A^{\frac{4}{3}} - x^{\frac{4}{3}}\right), \tag{6.2.24}$$

To calculate the period, we will integrate an expression derived from this equation over the interval from $t = 0$ to $t = T/4$. The corresponding interval in x is $x = A$ to $x = 0$. Further, over this interval

$$y = \frac{dx}{dt} < 0. \tag{6.2.25}$$

Therefore, we list the following set of mathematical manipulations

$$Y = \frac{dx}{dt} = -\sqrt{\frac{3}{2}}\left(A^{\frac{4}{3}} - x^{\frac{4}{3}}\right)^{\frac{1}{2}}, \tag{6.2.26}$$

$$\frac{dx}{\left(A^{\frac{4}{3}} - x^{\frac{4}{3}}\right)^{\frac{1}{2}}} = -\sqrt{\frac{3}{2}}\,dt, \tag{6.2.27}$$

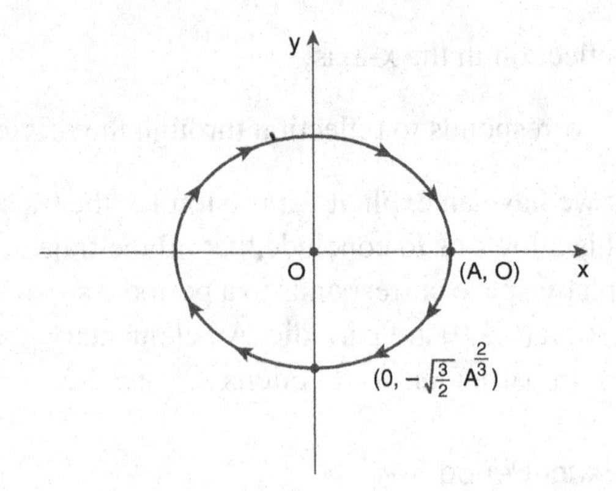

Figure 6.1: Trajectory curve in phase space for the initial conditions $x(0) = A$ and $y(0) = 0$.

and

$$\int_A^0 \frac{dx}{\sqrt{A^{\frac{4}{3}} - x^{\frac{4}{3}}}} = -\sqrt{\frac{3}{2}} \int_0^{\frac{T}{4}} dt. \tag{6.2.28}$$

Since

$$\int_A^0 (\cdots) = (-) \int_0^A (\cdots), \tag{6.2.29}$$

we have

$$\int_0^A \frac{dx}{\sqrt{A^{\frac{4}{3}} - x^{\frac{4}{3}}}} = \left(\frac{1}{4}\right) \sqrt{\frac{3}{2}} T. \tag{6.2.30}$$

Now make the following change of variables

$$x = Au, \tag{6.2.31}$$

and obtain

$$\begin{cases} X = 0 \Rightarrow u = 0, \\ X = A \Rightarrow u = 1, \\ dx = A du. \end{cases} \tag{6.2.32}$$

Therefore,

$$\int_0^A \frac{dx}{\sqrt{A^{\frac{4}{3}} - x^{\frac{4}{3}}}} = \int_0^1 \frac{A du}{\sqrt{A^{\frac{4}{3}}\left(1 - u^{\frac{4}{3}}\right)}} \tag{6.2.33}$$

$$= A^{\frac{1}{3}} \int_0^1 \frac{du}{\sqrt{1 - u^{\frac{4}{3}}}}.$$

The integral, on the right-side of this equation can be "simplified" by the change of variable

$$\begin{cases} v = u^{\frac{4}{3}} \text{ or } u = v^{\frac{3}{4}}, \\ du = \left(\frac{3}{4}\right) v^{-\frac{1}{4}} dv. \end{cases} \tag{6.2.34}$$

Making these substitutions on the right-side of Equation (6.2.33) and then using Equation (6.2.28), we find the period, T, the result

$$T = 3\left(\sqrt{\frac{2}{3}}\right) A^{\frac{1}{3}} \int_0^1 v^{-\frac{1}{4}}(1-v)^{-\frac{1}{2}} dv, \tag{6.2.35}$$

which can be written as

$$T = \phi\, A^{\frac{1}{3}}, \tag{6.2.36}$$

where the constant ϕ is

$$\phi = 3\left(\sqrt{\frac{2}{3}}\right) \int_0^1 v^{-\frac{1}{4}}(1-v)^{-\frac{1}{2}} dv. \tag{6.2.37}$$

In Equation (6.2.36), we have the exact period of the oscillation and it varies as the cube-root of the initial amplitude $x(0) = A$.

The integral in Equations (6.2.35) or (6.2.37) is a named definite integral and is called the Beta function. Its precise definition is

$$B(p, q) \equiv \int_0^1 v^{P-I}(I-v)^{q-1} dv. \tag{6.2.38}$$

For our special case

$$p - 1 = -\frac{1}{4}, \quad q - 1 = -\frac{1}{2}, \tag{6.2.39}$$

Consequently,

$$p = \frac{3}{4}, \quad q = \frac{1}{2}. \tag{6.2.40}$$

Therefore,

$$T = T(A) = 3\left(\frac{2}{3}\right)^{\frac{1}{2}} B\left(\frac{3}{4}, \frac{1}{2}\right) A^{\frac{1}{3}}. \tag{6.2.41}$$

There exists on the internet, websites that will calculate $B(0.75, 0.50)$; they are called Beta-calculators and return the value

$$B(0.75, 0.50) = 2.39628, \tag{6.2.42}$$

accurate to six significant places. Using this in Equation (6.2.41) gives for the period

$$T(A) = (5.86966)\, A^{\frac{1}{3}}, \tag{6.2.43}$$

or in terms of the angular velocity, Ω,

$$\Omega = \frac{2\pi}{T}, \tag{6.2.44}$$

the value

$$\Omega = \frac{1.07045}{A^{\frac{1}{3}}}. \tag{6.2.45}$$

6.3 APPROXIMATE SOLUTIONS

6.3.1 Preliminaries

The differential equation

$$\frac{d^2x}{dt^2} + x^{\frac{1}{3}} = 0, \tag{6.3.1}$$

is of odd parity, i.e., if $x \to -x$, then the resulting differential equation is the same except for an overall negative sign.

In general, if a system is modelled by an equation

$$F\left(\frac{d^2x}{dt^2}, \frac{dx}{dt}, x\right) = 0, \tag{6.3.2}$$

and if

$$F\left(-\frac{d^2x}{dt^2}, -\frac{dx}{dt}, -x\right) = -F\left(\frac{d^2x}{dt^2}, \frac{dx}{dt}, x\right), \tag{6.3.3}$$

then the system is said to have odd parity. Further, if the system is periodic and the solution, $x(t)$, can be expressed as a Fourier series, then only odd harmonics appear, i.e.,

$$x(t) = \sum_{k=0}^{\infty} \left[a_k \cos(2k+1)\Omega t + b_k \sin(2k+1)\Omega t\right], \tag{6.3.4}$$

where Ω is the angular frequency

$$\Omega = \frac{2\pi}{T}, \tag{6.3.5}$$

and T is the period.

Note that Equation (6.3.1) has no linear term in x. This means, as stated earlier, that none of the many standard perturbation techniques can be applied to it to determine approximations for its periodic solutions. Thus, new procedures must be created to obtain approximate solutions.

Oscillatory systems, such as Equation (6.3.1), are called "truly nonlinear oscillators" if they can be expressed as

$$\frac{d^2x}{dt^2} + f(x) = 0, \quad f(-x) = -f(x), \tag{6.3.6}$$

where $f(x)$ has no linear approximation at $x = 0$. Explicit examples are

$$f_1(x) = x^3, \quad f_2(x) = x^{\frac{1}{3}}, \quad f_3(x) = x + x^{\frac{1}{3}}. \tag{6.3.7}$$

Other explicit examples are

$$f_4(x) = |x|\,x, \; f_5(x) = \frac{x^3}{1+x^2}, \; f_6(x) = \frac{1}{x^{\frac{1}{3}}}. \tag{6.3.8}$$

If the periodic solutions to Equation (6.3.6) are represented as Fourier series, see Equation (6.3.4), then the coefficients satisfy the restrictions

$$|a_k| + |b_k| \leq \frac{M}{k^r}, \tag{6.3.9}$$

provided the first r-derivatives of $x(t)$ exist and where the positive constant may depend on $x(t)$ and T.

For the differential equation given by Equation (6.3.1) $f(x) = x^{\frac{1}{3}}$ and

$$\frac{df(x)}{dx} = \left(\frac{1}{3}\right)\left(\frac{1}{x^{\frac{2}{3}}}\right), \tag{6.3.10}$$

and the first derivative of $f(x)$ does not exist whenever $x = 0$. Also,

$$\begin{aligned} \frac{d^3x(t)}{dt^3} &= \frac{d}{dt}\left(\frac{d^2x(t)}{dt^2}\right) \\ &= -\left(\frac{1}{3}\right)\left(\frac{1}{x^{2/3}}\right)\frac{dx}{dt}. \end{aligned} \tag{6.3.11}$$

Therefore, the third time-derivative of $x(t)$ does not exist whenever $x = 0$. This means that $x(t)$ has a Taylor series expansion in the interval

$$|t| < \frac{T}{4}, \tag{6.3.12}$$

if we consider a solution where

$$x(0) = 1, \quad \frac{dx(0)}{dt} = 0. \tag{6.3.13}$$

(Please think about why this must be the case.)

Finally, let us examine Figure 6.1 where the dynamics, i.e., the behavior of the system as a function of time, can be thought of as a particle that moves along the closed phase space curve. The general features of this motion are as follows:

(a) The particle starts at $x(0) + A > 0, \; y(0) = 0$, and moves down the curve, until at time, $t = T/4$, it is at $x(T/4) = 0, \; y(T/4) = -\sqrt{\frac{3}{2}}A^{\frac{2}{3}}$.

(b) The particle then moves up the curve and at time, $t = T/2$, it is at $x(T/2) = -A, \; y(T/2) = 0$.

(c) At time, $t = 3T/4$, the particle is at $x(3T/4) = 0, \quad y(3T/4) = \sqrt{\frac{3}{2}}A^{\frac{2}{3}}$.

(d) Finally, at time $t = T$, the particle returns to its starting, initial point, and again begins its cyclic behavior.

(e) The time dependencies of $x(t)$ and $y(t)$ are similar to the general behaviors of the functions

$$A\cos\left(\frac{2\pi}{T}\right)t, \quad -A\sin\left(\frac{2\pi}{T}\right)t. \tag{6.3.14}$$

6.3.2 Harmonic Balance Approximation

Let us consider the following nonlinear differential equation, with indicated initial values:

$$\frac{d^2x(t)}{dt^2} + x(t)^{\frac{1}{3}} = 0; \; x(0) = A, \quad \frac{dx(0)}{dt} = 0. \tag{6.3.15}$$

The following results have either been established or can easily be shown to be true (see Figure 6.1):

(i) $x(t)$ is periodic with period T; see Equation (6.2.43).

(ii) $x(t)$ is an even function of t.

(iii) $x(t)$ is bounded by $A > 0$, i.e.,

$$-A \leq x(t) \leq A. \tag{6.3.16}$$

The above properties imply that $x(t)$ has a Fourier representation of the form

$$x(t) = \sum_{k=0}^{\infty} a_k \cos(2k+1)\;\Omega t, \tag{6.3.17}$$

where $\Omega = 2\pi/T$.

An approximate solution of order-N, in the **method of harmonic balance**, is given by an expression of the form

$$x_n(t) = \sum_{k=0}^{N-1} \bar{a}_k \cos(2k+1)\bar{\Omega}\,t, \tag{6.3.18}$$

where $\bar{a}_k$ and $\bar{\Omega}$ are a approximations to the actual Fourier coefficient, a_k, and the exact angular frequency, Ω. Consequently, we have

$$x_1(t) = A_1 \cos\bar{\Omega}\,t \tag{6.3.19}$$

$$x_2(t) = A_1 \cos\bar{\Omega}\,t + A_2 \cos 3\bar{\Omega}\,t, \tag{6.3.20}$$

where we have used a slightly changed notation for the amplitudes

To eliminate the one-third power term in the original differential equation, we cube Equation (6.3.15) and get

$$\left(\frac{d^2x}{dt^2}\right)^3 + x = 0. \tag{6.3.21}$$

Perhaps a better way of seeing this is to first write

$$\frac{d^2x}{dt^2} + x^{\frac{1}{3}} = 0 \tag{6.3.22}$$

as

$$\frac{d^2x}{dt^2} = -x^{\frac{1}{3}}, \tag{6.3.23}$$

and then cube both sides to obtain

$$\left(\frac{d^2x}{dt^2}\right)^3 = -x, \tag{6.3.24}$$

and then add x to both sides.

To obtain a first-order harmonic balance approximation, substitute Equation (6.3.19) into Equation (6.3.21) to obtain

$$\left[-\left(\bar{\Omega}\right)^2 A_1 \cos\theta\right]^3 + A_1 \cos\theta \simeq 0, \tag{6.3.25}$$

where $\theta = \bar{\Omega}t$. Using the trigonometric relationship

$$(\cos\theta)^3 = \left(\frac{3}{4}\right)\cos\theta + \left(\frac{1}{4}\right)\cos 3\theta, \tag{6.3.26}$$

we get

$$A_1\left[1 - \left(\frac{3}{4}\right)A_1^2\left(\bar{\Omega}\right)^6\right]\cos\theta + \text{HOH} \simeq 0, \tag{6.3.27}$$

where HOH contains higher-order contributions. Since our input approximation, $x_1 = A_1 \cos\theta$, only contain the function $\cos\theta$, we retain only that contribution for the calculation. This means that in Equation (6.3.27) the coefficient of the $\cos\theta$ term is zero. Since $A_1 = 0$ implies the trivial solution, we obtain

$$1 - \left(\frac{3}{4}\right)A_1^2\left(\bar{\Omega}\right)^6 = 0, \tag{6.3.28}$$

or

$$\bar{\Omega}(A_1) = \left(\frac{4}{3}\right)^{\frac{1}{6}} \frac{1}{A_1^{\frac{1}{3}}}. \tag{6.3.29}$$

Imposing the initial value $x_1(0) = A$, we obtain

$$A_1 = A, \tag{6.3.30}$$

and

$$\begin{aligned}\bar{\Omega}_1(A) &= \bar{\Omega}(A)\\ &= \frac{1.049115}{A^{\frac{1}{3}}},\end{aligned} \tag{6.3.31}$$

and this is to be compared with the exact value

$$\Omega_{\text{exact}}(A) = \frac{1.070451}{A^{\frac{1}{3}}}. \tag{6.3.32}$$

Thus, the first-order harmonic balance approximation to the solution is

$$x_1(t) = A\cos\left(\bar{\Omega}_1 t\right), \tag{6.3.33}$$

where $\bar{\Omega}_1$ is given by Equation (6.3.31).

A second-order approximation assumes that

$$x_2(t) = A_1\left[\cos\theta + z\cos 3\theta\right], \tag{6.3.34}$$

where

$$\theta = \bar{\Omega}_2\, t, \quad A_2 = zA_1. \tag{6.3.35}$$

Substituting Equation (6.3.34) into Equation (6.3.24), using

$$\begin{cases} (\cos\theta)^2 = \left(\frac{1}{2}\right) + \left(\frac{1}{2}\right)\cos 2\theta, \\ \cos\theta_1\cos\theta_2 = \left(\frac{1}{2}\right)\cos(\theta_1 - \theta_2) + \left(\frac{1}{2}\right)\cos(\theta_1 + \theta_2), \end{cases} \tag{6.3.36}$$

and simplifying the resulting expression, gives

$$\begin{aligned} &\left\{\left(\bar{\Omega}_2\right)^6 A_1^3\left[\left(\frac{3}{4}\right) + \left(\frac{27}{4}\right)z + \left(\frac{243}{2}\right)z^2\right] - A_1\right\}\cos\theta \\ &+ \left\{\left(\bar{\Omega}_2\right)^6 A_1^3\left[\left(\frac{1}{4}\right) + \left(\frac{27}{2}\right)z + \left(\frac{2187}{4}\right)z^3\right] - zA_1\right\}\cos 3\theta \\ &+ \text{HOH} \simeq 0. \end{aligned} \tag{6.3.37}$$

As before, the input into the harmonic balance procedure included only $\cos\theta$ and $\cos 3\theta$ terms, and we set these coefficients to zero in Equation (6.3.37) to give

$$\left(\bar{\Omega}_2\right)^6 A_1^2\left[\left(\frac{3}{4}\right) + \left(\frac{27}{4}\right)z + \left(\frac{243}{2}\right)z^2\right] = 1, \tag{6.3.38}$$

$$\left(\bar{\Omega}_2\right)^6 A_1^2\left[\left(\frac{1}{4}\right) + \left(\frac{27}{4}\right)z + \left(\frac{2187}{4}\right)z^3\right] = z. \tag{6.3.39}$$

Dividing these two expressions and simplifying gives

$$(1701)z^3 - (27)z^2 + 51z + 1 = 0. \tag{6.3.40}$$

Our interest is only in the smallest magnitude root of this equation and this turns out to be

$$z = -0.019178 \tag{6.3.41}$$

Question: Just looking at the linear part of the cubic equation, i.e.,

$$51z + 1 = 0, \tag{6.3.42}$$

gives

$$z = -\left(\frac{1}{51}\right) = -0.0196. \tag{6.3.43}$$

Note the closeness of the values of z in Equations (6.3.41) and (6.3.43). Can you explain why this must be so?

Imposing the requirement

$$x_2(0) = A_1(1 + z) = A, \tag{6.3.44}$$

gives

$$A_1 = \frac{A}{1 + z}, \tag{6.3.45}$$

which can then be used to determine $\bar{\Omega}_2$ from Equation (6.3.38). Doing this yields the expression

$$\begin{aligned} \bar{\Omega}_2 &= \left[\frac{1 + 2z + z^2}{\left(\frac{3}{4}\right) + \left(\frac{27}{4}\right) z + \left(\frac{243}{2}\right) z^2} \right] \left(\frac{1}{A^{\frac{1}{3}}} \right) \\ &= \frac{1.063410}{A^{\frac{1}{3}}}, \end{aligned} \tag{6.3.46}$$

and the second-order harmonic balance approximation

$$x_2(t) = \left(\frac{A}{1 + z} \right) [\cos \theta + z \cos 3\theta], \tag{6.3.47}$$

where $\theta = \bar{\Omega}_z t$.

Examination of Equation (6.3.47) shows that the ratio of the amplitudes of the first two-harmonics is

$$|z| = \left| \frac{A_2}{A_1} \right| \approx \frac{1}{51}, \tag{6.3.48}$$

Calculating to higher orders of approximation is consistent with this ratio holding for consecutive amplitudes, i.e.,

$$\left| \frac{A_{l+1}}{A_l} \right| = O\left(z^l\right), l = (1, 2, 3, \ldots). \tag{6.3.49}$$

Thus, the higher-order harmonics appear to decrease fast. However, this fact does not prove that the approximation to the Fourier series converges. The actual, exact Fourier series does converge and its coefficients are bounded by M/K^2, where M is a positive constant.

Let us assume that a measure of the “accuracy” of each order of the harmonic balance procedure is how close it gives for the associated angular frequency. For this measure, we use the percentage error, defined as

$$\text{Percentage Error (PE)} \equiv \left| \frac{\Omega_{\text{exact}} - \bar{\Omega}}{\Omega_{\text{exact}}} \right| \cdot 100. \tag{6.3.50}$$

Calculating the PE for the first and second harmonic balance approximations gives the values

$$PE(HB - 1) = 2.0\%, \quad PE(HB - 2) = 0.7\%. \tag{6.3.51}$$

Comment: The harmonic balance method involves only algebraic and elementary trigonometric operations. In general, the first- and second-order solutions may be calculated by hand. However, things become quite complex for higher-order approximations. For example, one way to do the third-order calculation is to select our approximation to be

$$\begin{aligned} x_3(\theta) &= A_1 \cos\theta + A_2 \cos 3\theta + A_3 \cos 5\theta \\ &= A_1 \left[\cos\theta + z_1 \cos 3\theta + z_2 \cos 5\theta\right], \end{aligned} \tag{6.3.52}$$

where

$$\theta = \bar{\Omega}_3 t, \quad z_1 = \frac{A_2}{A_1}, \quad z_2 = \frac{A_3}{A_2}, \tag{6.3.53}$$

The resulting algebraic equations for A_1, z_1 and z_2 are complex, cubic equations in two variables and are generally not easily solveable by hand.

The basis of the harmonic balance method is the assumption of a truncated trigonometric series for the solution, it substitution into the differential equation and the expansion of this result into another trigonometric series of the form

$$H_1(\cdots)\cos\,\theta + H_2(\cdots)\cos 3\,\theta + \cdots + H_N(\cdots)\cos(2N+1)\theta + HOH \simeq O. \tag{6.3.54}$$

Since the cosine functions, $\cos(2k+1)\theta$, are linearly independent, any linear combination of them that is equal to zero over an interval of θ must have the property that the various coefficients must be zero, i.e.,

$$H_k(\cdots) = 0, \quad k = (0, 1, \ldots, N). \tag{6.3.55}$$

6.3.3 Iteration Scheme

A major tool of mathematical scientists is linearization of nonlinear equations. We now show how this general methodology can be applied to constructing approximations to the periodic solutions of

$$\frac{d^2x}{dt^2} + x^{\frac{1}{3}} = 0; \quad x(0) = A, \quad \frac{dx(0)}{dt} = 0. \tag{6.3.56}$$

First, rewrite this differential equation to the form

$$x = -\left(\frac{d^2x}{dt^2}\right)^3, \tag{6.3.57}$$

and this corresponding to a cubic equation in the second derivative. One solution is given by the expression in Equation (6.3.56). However, it should be kept in mind that two other complex solutions exist and they are

$$\frac{d^2x}{dt^2} - \left(\frac{1+\sqrt{3}i}{2}\right)x^{\frac{1}{3}} = 0, \tag{6.3.58}$$

$$\frac{d^2x}{dt^2} - \left(\frac{1-\sqrt{3}i}{2}\right)x^{\frac{1}{3}} = 0. \tag{6.3.59}$$

This means that any approximation method based on Equation (6.3.57) will be influenced by the fact that this differential equation actually corresponds to three differential equations, namely, Equations (6.3.56), (6.3.58), and (6.3.59).

To proceed, multiply Equation (6.3.57) by a (for the moment, unknown) constant Ω^2, and add d^2x/dt^2 to both sides to obtain

$$\frac{d^2x}{dt^2} + \Omega^2 x = \frac{d^2x}{dt^2} - \Omega^2\left(\frac{d^2x}{dt^2}\right)^3. \tag{6.3.60}$$

Note that at this stage, we have only done several mathematic manipulations on Equation (6.3.57), so Equation (6.3.60) is just an alternative way of expressing Equation (6.3.57).

An iteration scheme is one for constructing a sequence of functions $x_k : k = (0, 1, 2, \ldots)$ such that given $x_o(t)$, we can calculate $x_1(t)$, and from this $x_2(t)$, etc. Each of the $x_k(t)$ is taken to be an approximation to the exact $x(t)$. For our purpose, we select the following scheme

$$\frac{d^2x_{k+1}}{dt^2} + \Omega_k^2 x_{k+1} = \frac{d^2x_k}{dt^2} - \Omega_k^2\left(\frac{d^2x_k}{dt^2}\right)^3, \tag{6.3.61}$$

where, Ω_k is to be determined, $x_0(t)$ is taken to be

$$x_0\,(t) = A\cos\left(\Omega_0 t\right), \tag{6.3.62}$$

and al the $x_k(t)$ satisfy the initial conditions

$$x_k\,(0) = A, \frac{dx_k(0)}{dt} = 0;\; k = (1,\ 2,\ 3, \ldots). \tag{6.3.63}$$

Of importance is the requirement that Ω_k is selected such that $x_{k+1}(t)$ be periodic.

Note that at the $(k + 1)$th stage of the calculation the functions

$$\{x_0\,(t),\ x,(t), \ldots, x_k\,(t)\}, \quad k = (1,\ 2,\ 3,\ \ldots), \tag{6.3.64}$$

will be known; therefore, the right-side of Equation (6.3.61) will be a given function of t. In other words, this differential equation is just a linear, second-order, inhomogeneous equation. Since all of the prior $x_k\,(t)$ are periodic and expressible in terms of cosine functions of $\theta = \Omega_k t$, Equation (6.3.61) can be easily solved by standard methods. An important result from using such techniques is that the solution, $x_{k+1}\,(t)$, will only be bounded, i.e., periodic, if the right-side of this equation does not contain a $\cos\theta$ term. Since Ω_k will appear in the coefficient of the $\cos\theta$ term, bounded solutions will exist if this coefficient is zero. It is from this condition that Ω_k is determined.

Returning to Equations (6.3.61) and (6.3.62), it follows that $x_1\,(t)$ satisfies the relation

$$\begin{aligned}\frac{d^2x_1}{dt^2} + \Omega_0^2 x_1 &= \left(-\Omega_0^2 A\cos\theta\right) - \left(-\Omega_0^2 A\cos\theta\right)^3 \\ &= -\left(\Omega_0^2\right)\left[1 - \Omega_0^6\left(\frac{3A^2}{4}\right)\right]A\cos\theta + \left(\frac{\Omega_0^8 A^3}{4}\right)\cos 3\theta.\end{aligned} \tag{6.3.65}$$

Setting the coefficient of the $\cos\theta$ term to zero gives.

$$1 - \Omega_0^6\left(\frac{3A^2}{4}\right) = 0, \tag{6.3.66}$$

and

$$\Omega_0(A) = \left(\frac{4}{3}\right)^{\frac{1}{6}} \left(\frac{1}{A^{1/3}}\right) \tag{6.3.67}$$

$$= \frac{1.0491151}{A^{1/3}}.$$

This is to be compared with the exact value

$$\Omega_{\text{exact}}(A) = \frac{1.070451}{A^{1/3}}. \tag{6.3.68}$$

Therefore, the percentage error (PE) is

$$PE(1) = \left|\frac{\Omega_{\text{exact}} - \Omega_0}{\Omega_{\text{exact}}}\right| \cdot 100 = 2\% \text{ error}. \tag{6.3.69}$$

The solution to

$$\frac{d^2x_1}{dt^2} + \Omega_0^2 x_1 = \left(\frac{\Omega_0^8 A^3}{4}\right) \cos 3\theta, \tag{6.3.70}$$

is

$$x_1(t) = A\left[\left(\frac{25}{24}\right)\cos(\Omega_0 t) - \left(\frac{1}{24}\right)\cos(3\Omega_0 t)\right], \tag{6.3.71}$$

Ω_0 is found in Equation (6.3.66). Note that the ratio of the two amplitudes is

$$\left|\frac{\frac{1}{24}}{\frac{25}{24}}\right| = \frac{1}{25} = 0.04, \tag{6.3.72}$$

is "small." This informs us that the dominant term in the Fourier expansion of the exact solution is (likely) the first term.

At the next level of the iteration, i.e., $k = 1$, we have

$$\frac{d^2x_2}{dt^2} + \Omega_1^2 x_2 = \frac{d^2x_1}{dt^2} - \Omega_1^2 \left(\frac{d^2x_1}{dt^2}\right)^3, \tag{6.3.73}$$

we must take $x_1(t)$ to be

$$\begin{cases} x_1(t) = A\left[\alpha\cos\theta - \beta\cos 3\theta\right], \\ \theta = \Omega_1 t, \quad \alpha = \dfrac{25}{24}, \quad \beta = \dfrac{1}{24}. \end{cases} \tag{6.3.74}$$

If this $x_1(t)$is substituted into the right-side of Equation (6.3.72), then the resulting expression obtained is

$$\frac{d^2x_2}{dt^2} + \Omega_1^2 x_2 = -\left(\Omega_1^2\right)\left[\alpha - \left(\frac{3A^2}{4}\right)\Omega_1^6\, h\right]\cos\theta + HOH, \tag{6.3.75}$$

where h is a function of α and β, i.e.,

$$h = h(\alpha, \beta) \equiv \left(\alpha^2 - \alpha\beta + 2\beta^2\right)\alpha. \tag{6.3.76}$$

the coefficient of $\cos\theta$ to zero allows the calculation of $\Omega_1(A)$; doing this gives

$$\begin{aligned}\Omega_1^6 &= \left(\frac{4}{3}\right)\left(\frac{1}{A^2}\right)\left[\frac{1}{\alpha^2 - \alpha\beta + 2\beta^2}\right] \\ &= \Omega_0^6\left[\frac{1}{\alpha^2 - \alpha\beta + 2\beta^2}\right],\end{aligned} \tag{6.3.77}$$

and

$$\Omega_1(A) = \frac{1.041424}{A^{1/3}}. \tag{6.3.78}$$

The corresponding percentage error is

$$PE(2) = 2.7\%, \tag{6.3.79}$$

which is slightly larger than the value given by the first integration calculation. This finding might be related to the fact that Equation (6.3.57), on which this iteration scheme is based, actually represents three different differential equations.

6.3.4 Comparison of the Harmonic Balance and Iteration Procedures

In general, for differential equations having periodic solutions, first-order calculations using the harmonic balance methods are easier to implement than iteration procedures. Of importance is the fact that harmonic balancing reduces the determination of approximate solutions to one of solving algebraic equations. However, these equations may become very complicated when condering the inclusion of higher-order harmonics.

Iterative methods reduce the original nonlinear differential equations to the solution of linear, second-order, inhomogeneous differential equations. While these equations may be directly integrated, the number of terms in the higher-order calculations may increase exponentially.

Which technique to apply for a given problem will depend on the particular system of interest.

Finally, it should be indicated that there are a variety of generalizations for both harmonic balance and iteration scheme. The book of Mickens (2010) discusses this along with certain related issues.

6.4 SUMMARY

This chapter was concerned with the question of constructing analytical approximations to the differential equation

$$\frac{d^2x}{dt^2} + x^{\frac{1}{3}} = 0: \ x(0) = A, \ \frac{dx(0)}{dt} = 0. \tag{6.4.1}$$

We showed that in the (x, y) phase space, where x and y are related by means of the system equations

$$\frac{dx}{dt} = y, \ \frac{dy}{dt} = -x^{\frac{1}{3}}, \tag{6.4.2}$$

the trajectories, $y = y(x)$, are solutions to the differential equation

$$\frac{dy}{dx} = -\frac{x^{\frac{1}{3}}}{y}, \tag{6.4.3}$$

and its integration gives the first-integral

$$H(x,y) = \frac{y^2}{2} + \left(\frac{3}{4}\right) x^{\frac{4}{3}} = \text{constant}. \tag{6.4.4}$$

The existence of this integral implies that all the trajectories in phase space are closed curves. Since closed curves in phase space correspond to periodic solutions, it follows that all solutions of Equation (6.4.1) are periodic.

Next, we demonstrated that the exact period can be explicitly calculated and its value is

$$T = \left(\frac{2\pi}{1.070451}\right) A^{\frac{1}{3}}. \tag{6.4.5}$$

Continuing with this effort, two schemes were proposed for constructing approximate solutions, namely, the method of harmonic balance and an iteration procedure. Both of these techniques allow for the calculation of approximate solutions. While only the lowest order solutions were determined, it was seen that the ratio of the amplitudes of the two lowest harmonics was small, i.e., for

$$x(t) = A_0 \cos\left(\bar{\Omega}\, t\right) + A_1 \cos\left(3\bar{\Omega}\, t\right) + \cdots, \tag{6.4.6}$$

$$\left|\frac{A_1}{A_0}\right| \approx 0.04. \tag{6.4.7}$$

This clearly implies, but does not prove, that the lowest order approximation is "accurate" to (about) four percent.

Examination of both the harmonic balance and iteration methodologies show that both are easy to implement and may be valueable tools to calculate analytic approximate solutions to other systems modelled by differential equations having periodic solutions.

PROBLEMS

Section 6.1

1) Do there exist physical systems for which either of the following differential equations could provide mathematical models?

$$\frac{d^2x}{dt^2} + x^3 = 0, \quad \frac{d^2x}{dt^2} + x^{\frac{1}{3}} = 0$$

Include in your considerations biological systems.

2) What is an exact solution to any type of equation?

Section 6.2

3) Why do closed paths in the (x, y) phase space correspond to periodic solutions? Do closed paths imply that the period is finite?

4) Consider the equation

$$\frac{d^2x}{dt^2} + x^{\frac{3}{5}} = 0; \quad x(0) = A, \quad \frac{dx(0)}{dt} = 0.$$

Are the solutions periodic? If so, can the period be calculated?

Section 6.3

5) An advanced form of harmonic balance is to assume that $x(t)$ can be approximated by the function

$$\bar{X}(t) = \frac{A_1 \cos\theta}{1 + B_1 \cos 2\theta},$$

where A_1 and B_1 are constants (to be determined), and $\theta = \Omega_1 t$, where Ω_1 is also to be determined. Use this function and apply it to

$$\frac{d^2x}{dt^2} + x^{\frac{1}{3}} = 0; \quad x(0) = A, \frac{dx(0)}{dt} = 0.$$

What is it necessary to have $\cos 2\theta$ in the denominator rather then $\cos\theta$?

NOTES AND REFERENCES

Sections 6.1 and 6.2: A general introduction to nonlinear oscillations that gives the basic fundaments of many of the standard perturbation techniques for differential equations having the form

$$\frac{d^2x}{dt^2} + x = \in f\left(x, \frac{dx}{dt}\right), 0 < \in \ll 1, \tag{6.4.8}$$

is the book

(1) R.E. Mickens, **Oscillations in Planor Dynamic Systems** (World Scientific, Singapore, 1996).

This book also discusses the method of harmonic balance and some of its applications (see Chapter 4), along with analysis of the properties of oscillatory systems in phase space (see Appendix I).

An elementary presentation on the Beta and Gamma functions and several of their applications is given in

(2) R.E. Mickens, **Mathematical Methods for the Natural and Engineering Sciences 2nd Edition** (World Scientific, London, 2017).

Section 4.4.6 also provides an introduction to the use of analyzing second-order differential equations in two-dimensional phase-space.

Section 6.3: Truly nonlinear oscillators have begun to be studied in detail in recent decades. The following book summaries many of the mathematical techniques used to construct approximations to their periodic solutions

(3) R.E. Mickens, **Truly Nonlinear Oscillators** (World Scientific, Singapore, 2010).

CHAPTER 7

Discretization of Differential Equations

7.1 INTRODUCTION

The general modelling methodology begins with a system that we would like to understand, followed by the construction of a set of mathematical equations modelling the system. In most cases, these equations are complex and no exact solutions are known. This issue is usually resolved by discretizing the equations. **Discretization** is the process of representing continuous variables, functions of these variables, operators, and equations by discrete counter-parts. This process is usually carried out as the initial step toward making these items suitable for numerical evaluation on digital computers.

A major difficulty with discretization is the existence of numerical instabilities, i.e., these may occur numerical solutions to the discrete equations that do not correspond to any solution of the original continuous equations.

As on illustrative example of this issue, we examine the decay differential equation

$$\frac{dx}{dt} = -\lambda x, \quad x(0) = x_0 > 0, \tag{7.1.1}$$

and discretize it using finite differences. For the discretization, we use

$$t \to t_k, \quad x(t) \to x_K. \tag{7.1.2}$$

$$\frac{dx}{dt} \to \frac{x_{k+1} - x_k}{h}, \tag{7.1.3}$$

$$h = \Delta t, \quad k = (0, 1, 2, 3, \ldots). \tag{7.1.4}$$

Note that the derivative is approximated by a forward-Euler (FE) discrete representation, which is based on the calculus definition of the first derivative

$$\frac{dx(t)}{dt} = \operatorname*{Lim}_{h \to 0} \frac{x(t+h) - x(t)}{h}. \tag{7.1.5}$$

DOI: 10.1201/9781003178972-7

A relationship that holds between the continuous and this discrete derivative is

$$\begin{aligned}\frac{dx(t)}{dt} &= \frac{x(t+h) - x(t)}{h} + O(h) \\ &= \frac{x_{k+1} - x_k}{h} + O(h)\end{aligned} \tag{7.1.6}$$

The existence of this and other similar relationships, such as

$$\begin{aligned}\frac{dx(t)}{dt} &= \frac{x(t) - x(t-h)}{h} + O(h) \\ &= \frac{x_k - x_{k-1}}{h} + O(h).\end{aligned} \tag{7.1.7}$$

and

$$\begin{aligned}\frac{dx(t)}{dt} &= \frac{x(t+h) - x(t-h)}{2h} + O\left(h^2\right) \\ &= \frac{x_{k+1} - x_{k-1}}{2h} + O\left(h^2\right),\end{aligned} \tag{7.1.8}$$

clearly imply that in general the discretization process is approximate. The discrete first derivatives, given in Equations (7.1.7) and (7.1.8), are known, respectively, as the backward-Euler (BE) and central-difference (CD) schemes.

Using the FE scheme, we have for the decay equation the following finite difference representation

$$\frac{x_{k+1} - x_k}{h} = -\lambda x_k, \quad x_0 = \text{given}, \tag{7.1.9}$$

where the $x(t)$ term has been replaced by x_k. Therefore,

$$x_{k+1} = (1 - \lambda h)x_k, \tag{7.1.10}$$

and it has the solution

$$x_k = x_0(1 - \lambda h)^k, \quad k = (1, 2, 3, \ldots), \tag{7.1.11}$$

observe that the solution to the differential equation is

$$x(t) = x_0 e^{-\lambda t}, \tag{7.1.12}$$

and monotonic decreases to zero for all $\lambda > 0$; further, it is also positive for $x_0 > 0$;. However, for x_k, there are five cases to study.

Case I: $0 < \lambda h \leq 1$

For this case,

$$0 < 1 - \lambda h < 1, \tag{7.1.13}$$

and x_k decreases monotonic to zero and, for finite k, is positive.

Case II: $\lambda h = 1$

Now we have

$$x_0 > 0; \quad x_k = 0, \text{ for } k = (1, 2, 3, \ldots), \tag{7.1.14}$$

i.e., except for $k = 0$, x_k is zero.

Case III: $1 - \lambda h < 2$

We have the two relations

$$1 - \lambda h < 0, \quad |1 - \lambda h| < 1, \tag{7.1.15}$$

i.e., x_k oscillates in sign, but decreases monotonic to zero.

Case IV: $\lambda h = 2$

The solution is

$$x_k = (-1)^k x_0, \tag{7.1.16}$$

with all the even $- k$ values x_0, while all the negative $-k$ values are $(-x_0)$.

Case V: $\lambda h > 0$

For this case, all solutions oscillate with a change of sign, in going from x_k to x_{k+1}. Further, the magnitude of x_k increases exponentially.

An examination of these five cases shows that only Case I generates numerical solutions that have all of the important properties of the solutions to the decay differential equation. The Cases II, III, IV, and V exhibit numerical instabilities.

If a BE scheme is constructed, i.e.,

$$\frac{x_k - x_{k-1}}{h} = -\lambda x_k, \quad x_0 = \text{given}, \tag{7.1.17}$$

then this can be rewritten as

$$x_{k+1} = \left(\frac{1}{1 + \lambda h}\right) x_k, \quad x_0 = \text{given}, \tag{7.1.18}$$

and his equation has the solution

$$x_k = x_0 \left(\frac{1}{1 + \lambda h}\right)^k, \tag{7.1.19}$$

which is positive and monotonic decreasing for all λh. Likewise, for a CD discretization, we have

$$\frac{x_{k+1} - x_{k-1}}{2h} = -\lambda x_k, \quad x_0 = \text{given} \tag{7.1.20}$$

and the solution

$$x_k = A(r_+)^k + B(r_-)^k, \tag{7.1.21}$$

with A and B being arbitrary constants. Since Equation (7.1.20) is a second-order, linear difference equation, two initial conditions are needed, x_0 and x_1. We are given x_0, and select x_1 to be, for example,

$$x_1 = \frac{x_0}{1 + \lambda h}. \tag{7.1.22}$$

The r_+ and r_- satisfy the restrictions

$$0 < r_+ < 1, \tag{7.1.23}$$

$$r_- < 0, \quad |r_-| > 1. \tag{7.1.24}$$

With this information, the following conclusions may be reached:

(i) The first term, on the right-side of the expression for x_k is bounded and decreases in magnitude to zero.

(ii) The second term oscillates in sign and has a magnitude that become unbounded as $k \to \infty$.

(iii) The CD scheme does not provide a valid discretization for the decay differential equation since all solutions become unbounded as $k \to \infty$.

Finally, with regard to the three above schemes, we have the results:

(a) The FE discretization produces a valid discretization if the step-size, $h = \Delta t$, satisfies the restriction

$$0 < h < \frac{1}{\lambda}. \tag{7.1.25}$$

(b) The BE scheme is valid for all positive values of h.

(c) The CD scheme does not provide a valid discretization for any positive h.

In summary, the example of the decay equation demonstrates that constructing valid or dynamic consistent finite difference discretizations for differential equations is a nontrivial task.

The purpose of this chapter is to introduce a "new" type of finite difference methodology for the discretization of differential equations. This set of procedures is called the nonstandard finite difference (NSFD) construction. It was created by Mickens in the late 1980's, first "formally" stated in 1995, and has been extensively applied to a broad range of differential equations; see Mickens (2005, 2021) and Patirdar (2016).

The nextion section introduces the concept of "exact finite difference schemes" and shows how to construct such discretizations for differential equations whose exact solution is known.

Section 7.3 presents the NSFD scheme methodology and introduces the concepts of denominator functions and nonlocal discrete representations of functions. The importance of positivity is emphasized for many of the equations to be discretized.

Sections 7.4 and 7.5 illustrate the construction of NSFD schemes for a broad set both ordinary and partial differential equations.

7.2 EXACT SCHEMES

7.2.1 Methodology

Consider a first-order scalar differential equation

$$\frac{dx}{dt} = f(x, t, p), \quad x(t_0) = x_0, \tag{7.2.1}$$

where $p = (p_1, p_2, \ldots, p_N)$ is a set of parameters used to characterize the system modelled by this equation. It is assumed that $f(x, t, p)$ is such that Equation (7.2.1) has a unique solution for $t > t_0$. Denote this solution by

$$x(t) = \phi(p, x_0, t_0, t), \tag{7.2.2}$$

with

$$\phi(p, x_0, t_0, t_0) = x_0. \tag{7.2.3}$$

Let a given finite difference discretization of Equation (7.2.1) be

$$x_{k+1} = g(p, h, x_k, t_k), \tag{7.2.4}$$

where $h = \Delta t,\ t_k = hk$, and $x_k \simeq x(t_k)$. Write the solution as

$$x_k = \Psi(p, h, x_0, t_0, t_k), \tag{7.2.5}$$

where the following condition holds

$$\Psi(p, h, x_0, t_0, t_0) = x_0. \tag{7.2.6}$$

Definition 7.1 Equations (7.2.1) and (7.2.4) are said to have the **same general solution** if and only if

$$x_k = x(t_k), \tag{7.2.7}$$

for all values of h for which x_k is defined.

Definition 7.2 An **exact finite difference scheme** is one for which the solution to the difference equation has the same general solution as the associated differential equation.

Theorem 7.1 The ordinary differential equation

$$\frac{dx}{dt} = f(x, t, p), \quad x(t_0) = x_0, \tag{7.2.8}$$

has the exact finite difference scheme

$$x_{k+1} = \phi(p, x_k,\ t_k,\ t_{k+1}) \tag{7.2.9}$$

In other words, given the general solution to Equation (7.2.7), see Equation (7.2.2), then the exact finite difference scheme is obtained by making the following substitutions

$$t_0 \to t_k, \quad t \to t_{k+1}, \quad x_0 \to x_k, \quad x(t) \to x_{k+1}. \tag{7.2.10}$$

in Equation (7.2.2).

For linear ordinary differential equations there is another technique for constructing exact difference schemes. Let $\left\{u^{(i)}(t) : i = 1, 2, \ldots, N\right\}$ be a set of N linearly, independent functions. Let

$$u_k^{(i)} \equiv u_k^{(i)}(t_k), \tag{7.2.11}$$

and form the determinant

$$\begin{vmatrix} u_k & u_k^{(1)} & \cdots & u_k^{(N)} \\ u_{k+1} & u_{k+1}^{(1)} & \cdots & u_{k+1}^{(N)} \\ \vdots & \vdots & & \vdots \\ u_{k+N} & u_{k+N}^{(1)} & \cdots & u_{k+N}^{(N)} \end{vmatrix} = 0. \tag{7.2.12}$$

Evaluation of this determinant will give an N-th order, linear ordinary difference equation and from this the exact scheme of the N-th order, ordinary differential equation, with solutions $\left\{u^{(i)}(t) : i = 1, 2, \ldots, N\right\}$ can be constructed.

Comment (Advanced Topic)

The above results allow us to understand the origin of numerical instabilities. To see this, note that the solutions $x(t)$ and x_k have the forms

$$x(t) = \phi(p, x_0, t_0, t), \qquad x_k = \psi(p, h, x_0, t_0, t_k), \tag{7.2.13}$$

where $x(t)$ depends on the N-parameters, p, while x_k depends on the $(N + 1)$ parameters, $p + h$. Thus, the continuous and discrete solutions lie in different dimensional spaces and this allows for bifurcations to occur.

For an introduction to bifurcation theory, presented without a lot of detailed mathematics, see S. Strogatz, **Nonlinear Dynamics and Chaos, 2nd Edition** (Westview Press; Boulder, Co; 2015). See Chapters 3 and 8.

7.2.2 Examples of Exact Schemes

This section gives eight examples of the explicit construction of exact finite difference schemes for differential equations whose general solutions are a priori known.

To begin, consider the **decay equation**

$$\frac{dx}{dt} = -\lambda x, \; x(t_0) = x_0, \tag{7.2.14}$$

having the solution

$$x(t) = x_0 e^{-\lambda(t-t_0)}. \tag{7.2.15}$$

Making the substitutions, stated in Equation (7.2.9), gives

$$x_{k+1} = e^{-\lambda h} x_k. \tag{7.2.16}$$

It follows that

$$\begin{aligned} x_{k+1} - x_k &= -\left(1 - e^{-\lambda h}\right) x_k \\ &= -\lambda \left(\frac{1 - e^{-\lambda h}}{\lambda}\right) x_k \end{aligned} \tag{7.2.17}$$

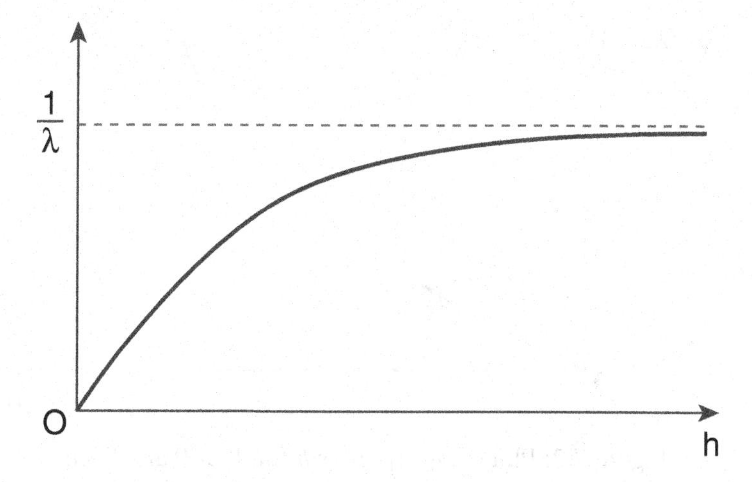

Figure 7.1: Plot of $\phi(\lambda, h)$ vs h for $\lambda > 0$ and fixed.

and, finally,

$$\frac{x_{k+1} - x_k}{\left(\frac{1-e^{-\lambda h}}{\lambda}\right)} = -\lambda x_k. \tag{7.2.18}$$

Observe that the derivative has the discretization

$$\frac{dx}{dt} \to \frac{x_{k+1} - x_k}{\phi(\lambda, h)}. \tag{7.2.19}$$

where $\phi(\lambda, h)$, called the **denominator function**, is given by

$$\phi(\lambda, h) = \frac{1 - e^{-\lambda h}}{\lambda}, \tag{7.2.20}$$

and has the properties

$$\phi(\lambda, h) = h + O\left(\lambda h^2\right), \tag{7.2.21}$$

$$\operatorname*{Lim}_{h\to\infty} \phi(\lambda, h) = \frac{1}{\lambda}, \tag{7.2.22}$$

$$\operatorname*{Lim}_{\lambda\to 0} \phi(\lambda, h) = h. \tag{7.2.23}$$

For fixed λ, $\phi(\lambda, h)$ has the geometric form shown in Figure 7.1.

The so-called **logistic differential equation** is the following nonlinear, first-order equation

$$\frac{dx}{dt} = \lambda_1 x - \lambda_2 x^2, \quad x(t_0) = x_0, \tag{7.2.24}$$

where (λ_1, λ_2) are non-negative parameters. This equation is separable and a solution can be found using partial fractions. This solution is

$$x(t) = \frac{\lambda_1 x_0}{(\lambda_1 - x_0\lambda_2)\exp\left[-\lambda_1 (t - t_0)\right] + \lambda_2 x_0}. \tag{7.2.25}$$

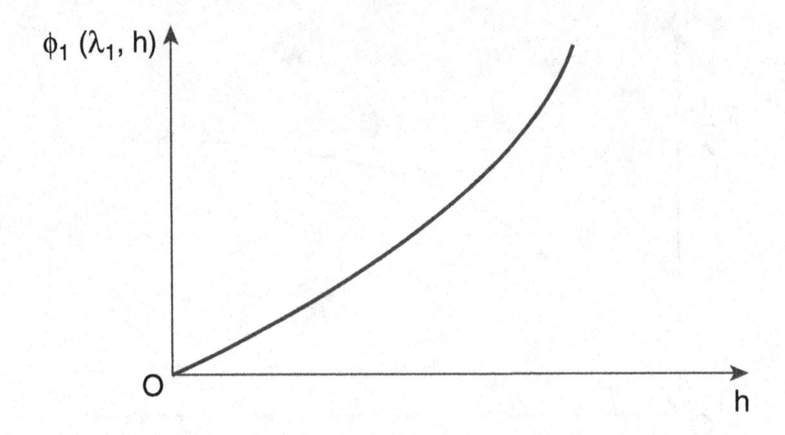

Figure 7.2: Plot of $\phi_1(\lambda_1, h)$ vs h for $\lambda_1 > 0$ and fixed.

Making the replacements, stated in Equation (7.2.9), the resulting expression can be rearranged to give

$$\frac{x_{k+1} - x_k}{\left(\frac{e^{\lambda_1 h} - 1}{\lambda_1}\right)} = \lambda_1 x_k - \lambda_2 x_{k+1} x_k. \tag{7.2.26}$$

Observe that

$$\frac{dx}{dt} \to \frac{x_{k+1} - x_k}{\left(\frac{e^{\lambda_1 h} - 1}{\lambda_1}\right)}, \qquad x^2 \to x_{k+1} x_k. \tag{7.2.27}$$

If, for this situation, we define the denominator function as

$$\phi_1(\lambda_1, h) = \frac{e^{\lambda_1 h} - 1}{\lambda_1}, \tag{7.2.28}$$

then it has the properties

$$\phi_1(\lambda_1, h) = h + O\left(\lambda_1 h^2\right) \tag{7.2.29}$$

$$\lim_{\lambda_1 \to 0} \phi_1(\lambda_1, h) = h, \tag{7.2.30}$$

$$\lim_{\lambda \to \infty} \phi_1(\lambda_1, h) = \infty. \tag{7.2.31}$$

These features are illustrated in Figure 7.2.

Of importance is the fact that the nonlinear term, x^2, is modelled nonlocally, i.e.,

$$x^2 \to x_{k+1} x_k, \tag{7.2.32}$$

and not

$$x^2 \not\to x_k^2 \tag{7.2.33}$$

The differential equation

$$\frac{dx}{dt} = -x^2, \quad x(t_0) = x_0, \tag{7.2.34}$$

has the solution

$$x(t) = \frac{x_0}{1 + x_0(t - t_0)}. \tag{7.2.35}$$

Making the substitution, given in Equation (7.2.9), produces

$$x_{k+1} = \frac{x_k}{1 + hx_K}, \tag{7.2.36}$$

and this can be rewritten to the form

$$\frac{x_{k+1} - x_k}{h} = -x_{k+1}x_k. \tag{7.2.37}$$

Note that the discrete derivative, for this case, is just the usual forward-Euler discretization

$$\frac{dx}{dt} \to \frac{x_{k+1} - x_k}{h}, \tag{7.2.38}$$

while the nonlinear x^2 term is modelled nonlocally; see Equation (7.2.30). This exact finite difference scheme can also be obtained by letting λ_1 go to zero in Equation (7.2.24) and setting $\lambda_2 = 1$.

For our fourth example, we consider the linear harmonic oscillator equation

$$\frac{d^2x}{dt^2} + \omega^2 x = 0, \tag{7.2.39}$$

where ω is the constant angular velocity. Its two linear independent solutions are

$$x^{(1)}(t) = e^{i\omega t}, \quad x^{(2)}(t) = e^{-i\omega t}, \tag{7.2.40}$$

where $i = \sqrt{-1}$. Using the determinant method, the exact scheme is calculated from

$$\begin{vmatrix} x_k & e^{i\omega hk} & e^{-i\omega hk} \\ x_{k+1} & e^{i\omega h(k+1)} & e^{-i\omega h(k+1)} \\ x_{k+2} & e^{(i\omega(k+2)} & e^{-i\omega h(k+2)} \end{vmatrix} = 0, \tag{7.2.41}$$

and this gives the following linear, second-order difference equation

$$x_{k+1} - [2\cos(\omega h)]x_k + x_{k-1} = 0, \tag{7.2.42}$$

Using the identity

$$2\cos\theta = 2 - 4\sin^2\left(\frac{\theta}{2}\right), \tag{7.2.43}$$

Equation (7.2.40) can be rewritten as

$$\frac{x_{k+1} - 2x_k + x_{k-1}}{\left(\frac{4}{\omega^2}\right)\sin^2\left(\frac{h\omega}{2}\right)} + \omega^2 x_k = 0, \tag{7.2.44}$$

and this is the exact finite difference scheme for the linear harmonic oscillator.

A particle of unit mass, acted on by a linear velocity force has the equation of motion

$$\frac{d^2x}{dt^2} = \lambda \frac{dx}{dt} \tag{7.2.45}$$

Its two linearly independent solutions are

$$x^{(1)}(t) = 1, \quad x^{(2)}(t) = e^{\lambda t}. \tag{7.2.46}$$

Therefore,

$$\begin{vmatrix} x_k & 1 & e^{\lambda hk} \\ x_{k+1} & 1 & e^{\lambda h(k+1)} \\ x_{k+2} & 1 & e^{\lambda h(k+2)} \end{vmatrix} = 0, \tag{7.2.47}$$

which, after same minor algebraic effort, can be expressed as

$$\frac{x_{k+1} - 2x_k + x_{k-1}}{\left(\frac{e^{\lambda h}-1}{\lambda}\right) h} = \lambda \left(\frac{x_k - x_{k-1}}{h}\right) \tag{7.2.48}$$

The one-dimension, unidirectional wave equation is, for $u = u(x, t)$,

$$\begin{cases} \dfrac{\partial u}{\partial t} + \dfrac{\partial u}{\partial x} = 0, \quad u(x, 0) = f(x) \text{ given,} \\ -\infty < x < \infty, \quad t > 0. \end{cases} \tag{7.2.49}$$

This first-order, linear partial differential equation has the solution

$$u(x, t) = f(x - t). \tag{7.2.50}$$

where it is assumed that $f(z)$ has a continuous first derivative. If we use the fact that the difference equation

$$u_m^{k+1} = u_{m-1}^k, \tag{7.2.51}$$

has the general solution

$$u_m^k = F(m - k), \tag{7.2.52}$$

then using

$$\Delta x = \Delta t, \tag{7.2.53}$$

and defining

$$\begin{cases} x_m = (\Delta x)m, \quad t_k = (\Delta t)k, \\ m = (\text{integers}), \\ k = (0, 1, 2, 3, \ldots), \end{cases} \tag{7.2.54}$$

it follows that under these conditions Equation (7.2.48) is the exact finite difference representation for the linear, unidirectional wave equation. Further, this expression can be rewritten to the form

$$\frac{u_m^{k+1} - u_m^k}{\phi(\Delta t)} + \frac{u_m^k - u_{m-1}^k}{\phi(\Delta x)} = 0, \quad \Delta x = \Delta t, \tag{7.2.55}$$

where $\phi(z)$ is any function satisfying the condition

$$\phi(z) = z + O\left(z^2\right). \tag{7.2.56}$$

A nonlinear version of the unidirectional wave equation is

$$\frac{\partial u}{\partial t} + \frac{\partial u}{\partial x} = u(1-u), \quad u(x,0) = f(x) \text{ given.} \tag{7.2.57}$$

The transformation

$$u(x,t) = \frac{1}{w(x,t)}, \tag{7.2.58}$$

gives the following linear, first-order, inhomogeneous equation

$$\frac{\partial w}{\partial t} + \frac{\partial w}{\partial x} = 1 - w, \quad w(x,0) = \frac{1}{f(x)}. \tag{7.2.59}$$

The solution to this equation is

$$w(x,t) = g(x-t)e^{-t} + 1, \tag{7.2.60}$$

where $g(z)$ is an arbitrary function of z having a first derivative in z. However, for $t = 0$, we have

$$w(x,0) = \frac{1}{f(x)} = g(x) + 1, \tag{7.2.61}$$

and we find that

$$g(x) = \frac{f(x) - 1}{f(x)} \tag{7.2.62}$$

Therefore, using the results from Equations (7.2.55), (7.2.57) and (7.2.59), we conclude that the solution to Equation (7.2.54) is

$$u(x,t) = \frac{f(x-t)}{e^{-t} + (1-e^{-t})\, f(x-t)}. \tag{7.2.63}$$

Solving for $f(x-t)$ gives

$$f(x-t) = \frac{e^{-t}u(x,t)}{1 - (1-e^{-t})\, u(x,t)}. \tag{7.2.64}$$

Making the substitutions

$$\begin{cases} x \to x_m = (\Delta x)m, \quad t \to t_k = (\Delta t)k, \\ u(x,t) \to u_m^k, \end{cases} \tag{7.2.65}$$

with

$$f(x,t) \to f[h(m-k)] = f_m^k, \quad \Delta x = \Delta t = h, \tag{7.2.66}$$

and using

$$f_m^{k+1} = f_{m-1}^k, \tag{7.2.67}$$

we obtain

$$\frac{e^{-h(k+1)} u_m^{k+1}}{1 - \left[1 - e^{-h(k+1)}\right] u_m^{k+1}} = \frac{e^{-hk} u_{m-1}^k}{1 - \left[1 - e^{-hk}\right] u_{m-1}^k}. \tag{7.2.68}$$

With some major algebraic work, the latter expression can be written as

$$\frac{u_m^{k+1} - u_m^k}{e^{\Delta t} - 1} + \frac{u_m^k - u_{m-1}^k}{e^{\Delta x} - 1} = u_{m-1}^k \left(1 - u_m^{k+1}\right), \tag{7.2.69}$$

with

$$\Delta x = \Delta t. \tag{7.2.70}$$

This is the exact scheme for the one-dimension, unidirectional wave equation with a $u(1-u)$ reaction term.

Finally, without providing details, we give the exact finite difference scheme for the second-order, linear differential equation

$$\frac{d^2\psi}{dr^2} + \left(\frac{2}{r}\right)\frac{d\psi}{dr} + \Omega^2 \psi = 0\,; \tag{7.2.71}$$

it is, using $r \to r_m = (\Delta r)m$,

$$\frac{\Psi_{m+1} - 2\psi_m + \Psi_{m-1}}{\left(\frac{4}{\Omega^2}\right)\sin^2\left(\frac{\Omega \Delta r}{2}\right)} + \left(\frac{2}{r_m}\right)\frac{\psi_{m+1} - \psi_{m-1}}{\left(\frac{2}{\Delta r}\right)\left(\frac{4}{\Omega^2}\right)\sin^2\left(\frac{\Omega \Delta r}{2}\right)} + \Omega^2 \psi_m = 0. \tag{7.2.72}$$

This equation, i.e., Equation (7.2.68), occurs in the analysis of wave motion having spherical
cal

symmetry.

An enhanced discussion of the derivation of these differential equations and several others in presented in Mickens (2021), Chapter 10.

7.3 NSFD METHODOLOGY

7.3.1 Important Ideas and Concepts

Based on a detailed examination of the exact finite difference structures determined from a large set of both ordinary and partial differential equations, Mickens formulated a new general methodology for constructing discretizations of differential equations. In concise, summary form they are:

(A) In general, discrete representations of the derivative have the structure

$$\frac{dx(t)}{dt} \to \frac{x_{k+1} - x_k}{\phi(p,h)}, \tag{7.3.1}$$

where p is the parameters appearing in the differential equation and $h = \Delta t$ is the time step-size. The $\phi(p, h)$, the **denominator function**, has the properties

$$\phi(p,h) = h + O\left(h^2\right), \quad \phi(0,h) = h, \tag{7.3.2}$$

$$\phi(p,h) > 0, \quad h > 0, \tag{7.3.3}$$

$$\frac{d\phi(p,h)}{dh} \geq 0, \quad h \geq 0. \tag{7.3.4}$$

(B) In general, functions of the dependent variable(s) should be **discretized nonlocally**. For
example,

$$\begin{aligned} x^2 &\rightarrow x_{k+1}\, x_k \\ &\not\rightarrow x_k^2. \end{aligned} \tag{7.3.5}$$

(C) **Dynamic consistency** between the differential equation(s) and the finite difference scheme(s) will generally allow both the denominator functions to be calculated and determine the mathematical structure of the discretizations of terms which are functions of the dependent variable(s).

An expected outcome of this methodology is to eliminate most, if not all, of the numerical instabilities which usually appear in the standard finite difference schemes.

A powerful tool in the nonstandard finite difference (NSFD) construction process is the use of sub-equations. A **sub-equation** is defined as follows: Let a differential equation be compossed of $N \geq 3$ separate terms. A sub-equation is any differential equation formed by including any $(N - 1)$ or fewer terms of the original differential equation. For example, consider, $u = u(x, t)$,

$$\frac{\partial u}{\partial t} + a\frac{\partial u}{\partial x} = D\frac{\partial^2 u}{\partial x^2} + u(1-u), \tag{7.3.6}$$

then some of the sub-equations are

$$\frac{\partial u}{\partial t} + a\frac{\partial u}{\partial x} = 0, \tag{7.3.7}$$

$$a\frac{\partial u}{\partial x} = u(1-u), \tag{7.3.8}$$

$$\frac{\partial u}{\partial t} \neq a\frac{\partial u}{\partial x} = u(1-u), \tag{7.3.9}$$

$$\frac{\partial u}{\partial t} = D\frac{\partial^2 u}{\partial x^2}, \tag{7.3.10}$$

$$D\frac{\partial^2 u}{\partial x^2} + u(1-u) = 0, \tag{7.3.11}$$

$$a\frac{\partial u}{\partial x} = D\frac{\partial^2 u}{\partial x^2}. \tag{7.3.12}$$

Observe that Equations (7.3.8), (7.3.11) and (7.3.12) are ordinary differential equations, while the other equations are partial differential equations. Also, of great interest and value is the fact that Equations (7.3.7), (7.3.8), (7.3.9) and (7.3.12) have known exact finite difference schemes.

Note that Equations (7.3.9) and (7.3.12) both contain the term "$a\ \partial u/\partial x$." Consequently, a NSFD scheme can be constructed for Equation (7.3.6) by combining the two exact discretizations such that the "$a\ \partial u/\partial x$" term has the same discretization. This process can be extended to the general construction of NSFD schemes for differential equations as will be demonstrated in the next two sections.

Finally, it should be stressed that NSFD schemes are not exact schemes, although after the fact, this may turn out to be the case. The major hope is that this methodology provides "better" discretizations than these provided by the standard techniques.

7.3.2 Constructing NSFD Schemes

How does one go about constructing a NSFD scheme for a particular set of differential equations? We provide below a general set of steps which can be used for this task. However, it must be kept in mind that these steps are only suggestive of what may be done, since there can be no a priori formulated deterministic methods for this activity.

The nine steps are:

(1) Start with the given differential equations and make certain that they are both "physically" and "mathematically" correct/valid.

(2) Study the equations to determine what parameters appear, their signs, relative magnitudes.

(3) Determine the physical scales of both the dependent and independent variables. Use these to form dimensionless equations. The values of the initial and/or boundary conditions may play important roles at this stage of the investigation.

(4) Investigate, in as much detail as possible, both the qualitative and quantitative properties of the solutions to the differential equations.

(5) From Step-4, determine which essential features that you wish to have incorporated into the finite difference discretization.

(6) Examine relevant sub-equations and determine the conditions under which their solutions provide additional insights into the behavior of the solutions to the full set of differential equations.

(7) Using the method of sub-equations, discretize the relevant sub-equations and learn what possibilities exist for both the denominator functions and the non-local representations for terms that depend on the independent variables.

(8) Check this discretization of the full set of eqations to verify that it is dynamically consistent with all features that you consider to be important, as given in Step-5.

(9) If necessary, repeat the full cycle.

Finally, it should be clear that the NSFD methodology does not provide a unique NSFD scheme for a set of differential equations. For example, different groups of sub-equations will yield different NSFD schemes. However, the more insights that the modeler has into the origins and uses of the differential equations, the better will be the overall quality of the discretization.

7.4 NSFD SCHEMES FOR ONE'S

This section presents five sets of differential equations derived from modelling phenomena in a broad rangė of the natural sciences. We show how such equations may be discretized in terms of the NSFD methodology. Additional examples and enhanced discussions are given in Mickens (2005).

7.4.1 Modified Newton Law of Cooling

A generalization of Newton's law of cooling is the equation

$$\frac{dx}{dt} = -\lambda x - \epsilon x^{\frac{1}{3}}, \quad x(0) = x_0, \tag{7.4.1}$$

where λ and $\in$ are non-negative parameters. The critical sub-equation here is

$$\frac{dx}{dt} = -\lambda x, \tag{7.4.2}$$

and it has the exact finite difference scheme

$$\frac{x_{k+1} - x_k}{\phi(\lambda, h)} = -\lambda\, x_{k+1}, \tag{7.4.3}$$

where

$$\phi(\lambda, h) = \frac{e^{\lambda h} - 1}{\lambda}. \tag{7.4.4}$$

Consequently, we are seeking a NSFD scheme having the form

$$\frac{x_{k+1} - x_k}{\phi} = -\lambda x_{k+1} - \epsilon \left[x^{\frac{1}{3}}\right]_k, \tag{7.4.5}$$

and we have to select an appropriate discretization for the term in the square brackets.

If, $x_0 > 0$, then the solution to Equation (7.4.1) has the following properties:

(i) $x(t) > 0, \quad t > 0;$

(ii) $\frac{dx(t)}{dt} < 0, \quad t > 0;$

(iii) $\lim_{t \to \infty} x(t) = 0.$

In other words, $x(t)$ decreases smoothly through positive values to the value zero. One way to enforce the positivity of the corresponding x_k is to use the following ansatz

$$\left[x^{\frac{1}{3}}\right]_k = \frac{x_{k+1}}{(x_k)^{\frac{2}{3}}}. \tag{7.4.6}$$

Note that

$$\lim_{\substack{h\to 0\\ k\to\infty\\ t=hk=\text{fixed}}} \left[x^{\frac{1}{3}}\right]_k = x^{\frac{1}{3}}. \tag{7.4.7}$$

Combining Equations (7.4.5) and (7.4.6), we obtain the following NSFD discretization for Equation (7.4.1)

$$\frac{x_{k+1} - x_k}{\phi} = -\lambda x_{k+1} - \in \frac{x_{k+1}}{(x_k)^{2/3}}. \tag{7.4.8}$$

This equation is linear in x_{k+1} and thus can be solved for it. Also, using the definition of ϕ, from Equation (7.4.4), we have

$$1 + \lambda\phi = e^{\lambda h}, \tag{7.4.9}$$

and

$$x_{k+1} = \left[\frac{(x_k)^{\frac{2}{3}}}{e^{\lambda h}(x_k)^{\frac{2}{3}} + \epsilon}\right] x_k\,. \tag{7.4.10}$$

Since,

$$0 < \frac{(x_k)^{\frac{2}{3}}}{e^{\lambda h}(x_k)^{2/3} + \epsilon} < 1, \tag{7.4.11}$$

it follows that the discrete solution has the following properties:

(a) $x_0 > 0 \Rightarrow x_k > 0$;

(b) $x_{k+1} < x_k$;

(c) $\lim_{k\to\infty} x_k = 0$.

Comparison of (i), (ii) and (iii), with (a), (b) and (c), allows for the conclusion that this NSFD discretization is dynamically consistent, with respect to these three properties, to the differential equation.

Finally, observe that this NSFD scheme is explicit, i.e., given x_k, the value of x_{k+1} can be directly calculated.

7.4.2 Stellar Structure

An understanding of the structure of star evolution is fundamental to our comprehension of the overall properties of the universe. The basic equations to describe a spherically symmetric star depend on the following variables:

$$\begin{aligned} P(r) &= \text{pressure} \\ M(r) &= \text{mass} \\ L(r) &= \text{luminosity} \\ \rho(r) &= \text{mass density,} \end{aligned}$$

where r is the radial distance from the center of the star and $F(r)$ is the value of property, "F," at r.

The density $\rho(r)$ can be gotten from the equation of state for an idea gas

$$\rho(r) = \frac{P(r)}{KT(r)}, \tag{7.4.12}$$

where K is a known, positive constant, and T is the temperature. Denote by ϵ, the non-negative energy production function which depends on ρ and T, i.e.,

$$\epsilon = \epsilon(\rho, T) \geq 0. \tag{7.4.13}$$

We assume that $\epsilon(\rho, T)$ is given. From these quantities, the following differential equations allow for the calculation of the pressure, mass, luminosity, and temperature as functions of r:

$$\frac{dP}{dr} = -\left(\frac{GM\rho}{r^2}\right), \tag{7.4.14}$$

$$\frac{dM}{dr} = 4\pi^2\rho, \tag{7.4.15}$$

$$\frac{dL}{dr} = 4\pi r^2 \rho\, \epsilon, \tag{7.4.16}$$

$$\frac{dT}{dt} = -\left(\frac{\gamma - 1}{\gamma}\right)\left(\frac{GMT}{r^2}\right), \tag{7.4.17}$$

where G is the Newton gravitational constant and γ is a constant greater than one.

These four variables, (P, M, L, T) have the following properties:

(i) $P(r) \geq 0, \;\; r > 0,$
$M(r) \geq 0, \;\; r > 0,$
$L(r) \geq 0, \;\; r > 0,$
$T(r) \geq 0, \;\; r > 0.$

(ii) $P(r)$ and $T(r)$, both decrease monotonic from the center of the star to its surface.

(iii) $M(r)$ and $L(r)$, both increase monotonic from the center of the star to its surface.

We now show how to construct NSFD schemes such that the discretized variables have the above three properties. The following notation will be used

$$r \to r_m = (\Delta r)m, \;\; \Delta r = h, \;\; m = (0, 1, 2, \ldots); \tag{7.4.18}$$

$$\begin{pmatrix} P(r) \\ M(r) \\ L(r) \\ T(r) \end{pmatrix} \to \begin{pmatrix} P_m \\ M_m \\ L_m \\ T_m \end{pmatrix} \tag{7.4.19}$$

To begin, write Equation (7.4.14) as

$$\frac{dP}{dt} = -\left(\frac{G}{K}\right)\left(\frac{MP}{r^2 T}\right), \tag{7.4.20}$$

where ρ is replaced by the expression in Equation (7.4.12). It follows that

$$\frac{dP}{dt} = -2\left(\frac{G}{K}\right)\left(\frac{MP}{r^2T}\right) + \left(\frac{G}{K}\right)\left(\frac{MP}{r^2T}\right) \tag{7.4.21}$$

and we take the NSFD discretization to be

$$\frac{P_{m+1} - P_m}{h} = -2\left(\frac{G}{K}\right)\left(\frac{M_m P_{m+1}}{r_m^2 T_m}\right) + \left(\frac{G}{K}\right)\left(\frac{M_m P_m}{r_m^2 T_m}\right). \tag{7.4.22}$$

Solving for P_{m+1} gives

$$P_{m+1} = \left[\frac{r_m^2 T_m + \left(\frac{hG}{K}\right) M_m}{r_m^2 T_m + 2\left(\frac{hG}{K}\right) M_m}\right] P_m. \tag{7.4.23}$$

Note that for

$$T_m \geq 0, \ M_m \geq 0, \ P_m \geq 0, \tag{7.4.24}$$

then

$$0 < P_{m+1} < P_m. \tag{7.4.25}$$

Next, replace ρ in Equation (7.4.15) by its value in Equation (7.4.12), to get

$$\frac{dM}{dt} = \left(\frac{4\pi}{K}\right)\left(\frac{r^2 P}{T}\right), \tag{7.4.26}$$

and discretize it as follows

$$\frac{M_{m+1} - M_m}{h} = \left(\frac{4\pi}{K}\right)\left(\frac{r_m^2 P_{m+1}}{T_m}\right). \tag{7.4.27}$$

Solving for M_{m+1} gives

$$M_{m+1} = M_m + \left(\frac{4\pi h}{K}\right)\left(\frac{r_m^2 P_{m+1}}{T_m}\right). \tag{7.4.28}$$

Since we know P_{m+1} from Equation (7.4.23), then Equation (7.4.28) allows M_m to be determined. Another consequence is

$$M_{m+1} > M_m > 0, M_0 > 0. \tag{7.4.29}$$

Now rewrite the energy production function as

$$\epsilon(\rho, T) = \epsilon\left(\frac{P}{KT}, T\right), \tag{7.4.30}$$

and define the function $E(\rho, T)$ as

$$\begin{aligned} E(\rho, T) &\equiv \rho\epsilon \\ &= \left(\frac{P}{KT}\right) \in \left(\frac{P}{KT}, T\right) \\ &= E_1(P, T). \end{aligned} \tag{7.4.31}$$

With this definition, the differential equation for $L(r)$ becomes

$$\frac{dL}{dr} = 4\pi r^2 E_1(P, T)_r, \tag{7.4.32}$$

and we select the NSFD discretization to be

$$\frac{L_{m+1} - L_m}{h} = 4\pi T_m^2 E_1 (P_{m+1}, T_m). \tag{7.4.33}$$

Solving for L_{m+1} gives

$$L_{m+1} = L_m + 4\pi r_m^2 E_1 (P_{m+1}, T_m), \tag{7.4.34}$$

with the following conditions holding

$$L_{m+1} > L_m > 0, \quad L_0 > 0. \tag{7.4.35}$$

For the dT/dr equation, define β to be

$$\beta \equiv \left(\frac{\gamma - 1}{\gamma}\right) G, \tag{7.4.36}$$

and write the resulting differential equation as

$$\frac{dT}{dt} = -2\beta\left(\frac{MT}{r^2}\right) + \beta\left(\frac{MT}{r^2}\right). \tag{7.4.37}$$

We take the following as its NSFD discretization

$$\frac{T_{m+1} - T_m}{h} = -2\beta\left(\frac{M_{m+1}T_{m+1}}{r_m^2}\right) + \beta\left(\frac{M_{m+1}T_m}{r_m^2}\right). \tag{7.4.38}$$

Solving for T_{m+1} gives

$$T_{m+1} = \left[\frac{r_m^2 + (h\beta)M_{m+1}}{r_m^2 + 2(h\beta)M_{m+1}}\right] T_m, \tag{7.4.39}$$

from which it follows that

$$0 < T_{m+1} < T_m. \tag{7.4.40}$$

In summary, we have shown how to construct a NSFD scheme for the differential equations modelling stellar structure. Note that we have, in several cases, written an expression y as

$$y = 2y - y \rightarrow 2y_m - y_{m+1}, \tag{7.4.41}$$

to achieve a positivity preserving scheme. Further, the denominator functions for all of the discretized equation is $\phi(h) = h$ and not a more complex form. Also, the NSFD scheme is dynamic consistent with respect to all the important properties of the solutions to the differential equations. Finally, observe that this scheme is sequential, in that their numerical values must be determined in a specific order, i.e.,

(i) From (r_m, M_m, P_m), the value P_{m+1} is calculated.

(ii) Next, M_{m+1} is calculated from knowledge of (P_{m+1}, T_m, M_m).

(iii) L_{m+1} is determined from (L_m, T_m, P_{m+1}).

(iv) Finally, T_{m+1} is calculated using (T_m, M_{m+1}).

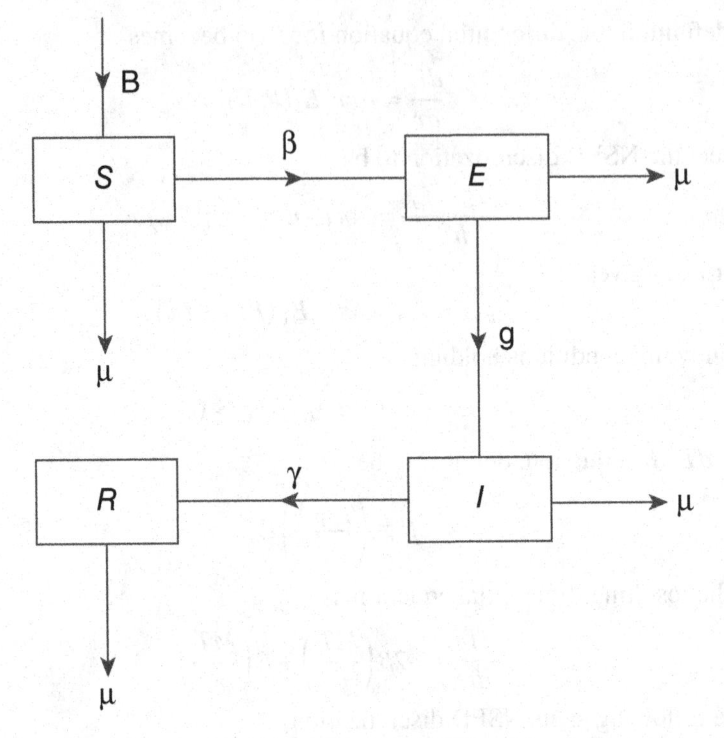

Figure 7.3: Compartmental diagram for an SEIR epidemiological model.

7.4.3 SEIR Model

An important mathematical epidemiological model for the spread of disease is the SEIR formulation. The four dependent variables are

$S(t)$ – susceptible population
$E(t)$ – infected population
$I(t)$ – infectious population
$R(t)$ – recovered/removed population.

If compartments are assigned to each population, then Figure 7.3 shows how the populations move through the system.

The SEIR model is represented by four coupled, first-order, ordinary differential equations:

$$\frac{dS}{dt} = B - \mu S - \beta\left(\frac{I}{N}\right)S, \tag{7.4.42}$$

$$\frac{dE}{dt} = \beta\left(\frac{I}{N}\right)S - gE - \mu E, \tag{7.4.43}$$

$$\frac{dI}{dt} = gE - \mu I - \gamma I, \tag{7.4.44}$$

$$\frac{dR}{dt} = \gamma I - \mu R, \tag{7.4.45}$$

where the parameters $(B, g, \mu, \beta, \gamma)$ are non-negative, and $N(t)$ is the total population, i.e.,

$$N(t) = S(t) + E(t) + I(t) + R(t). \tag{7.4.46}$$

Adding the above four differential equations and using the definition of $N(t)$, we obtain the generalized **conservation law**

$$\frac{dN(t)}{dt} = B - \mu N(t), \tag{7.4.47}$$

whose exact solution is

$$N(t) = \left(N_0 - \frac{B}{\mu}\right) e^{-\mu t} + \frac{B}{\mu}, \tag{7.4.48}$$

where

$$N_0 = S(0) + E(0) + I(0) + R(0). \tag{7.4.49}$$

The exact difference scheme is

$$\frac{N_{k+1} - N_k}{\phi} = B - \mu N_k, \tag{7.4.50}$$

$$\phi = \frac{1 - e^{-\mu h}}{\mu}. \tag{7.4.51}$$

Since conservation laws are critical for the evolution of a system, it is important to incorporate them into any NSFD scheme. For this system, this can be done by requiring, each of the derivative discretizations to have the same denominator function, namely, ϕ as given in Equation (7.4.51). A NSFD scheme which satisfies positivity, i.e.,

$$\begin{pmatrix} S_k \\ E_k \\ I_k \\ R_k \end{pmatrix} \geq 0 \Rightarrow \begin{pmatrix} S_{k+1} \\ E_{k+1} \\ I_{k+1} \\ R_{k+1} \end{pmatrix} \geq 0, \tag{7.4.52}$$

is

$$\frac{S_{k+1} - S_k}{\phi} = B - \mu S_k - \beta\left(\frac{I_k}{N_k}\right) S_{k+1}, \tag{7.4.53}$$

$$\frac{E_{k+1} - E_k}{\phi} = \beta\left(\frac{I_k}{N_k}\right) S_{k+1} - gE_{k+1} - \mu E_k, \tag{7.4.54}$$

$$\frac{I_{k+1} - I_k}{\phi} = gE_{k+1} - \mu I_k - \gamma I_{k+1}, \tag{7.4.55}$$

$$\frac{R_{k+1} - R_k}{\phi} = \gamma I_{k+1} - \mu R_k. \tag{7.4.56}$$

Several observations can be made. First, if the last four equations are added, then we obtain the result of Equation (7.4.50). This means that while the individual discretizations are not exact, the collective set of discretizations satisfies exactly the conservation law.

Second, we can solve these equations for the variables at the $(k+1)$th discrete time level to obtain the expressions

$$S_{k+1} = \frac{B\phi + \left(e^{-\mu h}\right) S_k}{1 + (\beta\phi)\left(\frac{I_k}{N_k}\right)}, \tag{7.4.57}$$

$$E_{k+1} = \frac{\left(e^{-\mu h}\right) E_k + (\beta\phi) S_{k+1}\left(\frac{I_k}{N_k}\right)}{1 + g\phi}. \tag{7.4.58}$$

$$I_{k+1} = \frac{\left(e^{-\mu h}\right) I_k + (g\phi) E_{k+1}}{1 + \gamma\phi}, \tag{7.4.59}$$

$$R_{k+1} = (\gamma\phi) I_{k+1} + \left(e^{-\gamma h}\right) R_k. \tag{7.4.60}$$

The $(S_{k+1}, E_{k+1}, I_{k+1}, R_{k+1})$ are to be calculated in the following order

(a) first, calculated S_{k+1} from knowledge of (S_k, E_k, I_k, R_k);

(b) second, calculate E_{k+1} from $(S_{k+1}, S_k,\ E_k,\ I_k,\ R_k)$;

(c) third, calculate I_{k+1} from $(I_k,\ E_{k+1})$;

(d) fourth, obtain R_{k+1} from $(I_{k+1},\ R_k)$.

Inspection of Equations (7.4.57) to (7.4.60) shows that the positivity requirement holds.

7.4.4 Time-Independent Schrödinger Equations

The following differential equation

$$\frac{d^2u}{dx^2} + f(x)u = 0, \tag{7.4.61}$$

where $f(x)$ is specified, is called a time-independent Schrödinger equation because it can be derived from the Schrödinger equation, i.e.,

$$-\left(\frac{\hbar^2}{2m}\right)\phi''(x) + V(x)\phi(x) = E\phi(x), \tag{7.4.62}$$

is the fundamental equation describing the quantum behavior of a single particle of mass m, inter-acting with a potential $V(x)$. E is the total energy and $\hbar = h/2\pi$ is the reduced Planck constant. Also, $\phi''(x) = d^2\phi(x) \neq dx^2$. This equation can be rewritten to the form given in Equation (7.4.61) if

$$\phi(x) \to u(x), \left(\frac{2m}{\hbar^2}\right)[E - V(x)] \to f(x). \tag{7.4.63}$$

If we use the exact finite difference scheme for

$$\frac{d^2u}{dx^2} + \lambda u = 0, \quad \lambda = \text{constant}, \tag{7.4.64}$$

i.e.,

$$\frac{u_{m+1} - 2u_m + u_{m-1}}{\left(\frac{4}{\lambda}\right)\sin^2\left(\frac{h\sqrt{\lambda}}{2}\right)} + \lambda u_m = 0, \tag{7.4.65}$$

and replace

$$\lambda \to f(x_m) = f_m, \tag{7.4.66}$$

then a NSFD scheme for Equation (7.4.61) is

$$\frac{u_{m+1} - 2u_m + u_{m-1}}{\left(\frac{4}{f_m}\right)\sin^2\left(\frac{h\sqrt{f_m}}{2}\right)} + f_m u_m = 0, \tag{7.4.67}$$

where

$$\Delta x = h, \ x \to x_m = (\Delta x)m, \ u(x) \to u_m. \tag{7.4.68}$$

Note that if use is made of the trigonometric relation

$$\sin(i\theta) = i\sinh(\theta), \quad i = \sqrt{-1}, \tag{7.4.69}$$

then Equation (7.4.67) holds for positive and negative values of $f_m = f(x_m)$.

Using the identity

$$2(\sin\theta)^2 = 1 - \cos(2\theta), \tag{7.4.70}$$

Equation (7.4.67) can be rewritten as

$$u_{m+1} + u_{m-1} = 2\left[\cos\left(h\sqrt{f_m}\right)\right]u_m. \tag{7.4.71}$$

7.4.5 Modified Anderson-May Model

This section deals with issues that are different from those previously considered in this section. Our task is to convert a particular discrete model into the corresponding differential equations. The original discrete model is the Anderson-May representation for the spread of a (periodic) disease. The relevant references and extensions of the work shown here is given is Section 11.9.2 of Mickens (2021).

The following pair of first-order difference equations were given by *R.* Anderson and *R.* May to describe the interactions of susceptible and infective populations,

$$S_{k+1} = B + S_k - \beta S_k I_k, \tag{7.4.72}$$

$$I_{k+1} = \beta S_k I_k, \tag{7.4.73}$$

where B and β are non-negative parameters. These two coupled difference equations do not have a corresponding set of associated differential equations. (As an interesting exercise, try to find two coupled, first-order differential equations that have the above difference equations as their discretizations.) However, if Equations (7.4.72) and (7.4.73) are modified, then differential equations will exist.

To show this, note that

$$S_k - \beta S_k I_k = (1 - \beta I_k) S_k, \tag{7.4.74}$$

and for "small" β we have

$$(1 - \beta I_k) S_k \rightarrow S_k e^{-\beta I_k}, \tag{7.4.75}$$

$$\beta S_k I_k \rightarrow S_k \left(1 - e^{-\beta I_k}\right). \tag{7.4.76}$$

Making these substitutions in Equations (7.4.72) and (7.4.73) gives the modified Anderson-May equations

$$S_{k+1} = B + S_k e^{-\beta I_k}, \tag{7.4.77}$$

$$I_{k+1} = S_k \left(1 - e^{-\beta I_k}\right) + e^{-a} I_k, \tag{7.4.78}$$

$$R_{k+1} = (1 - e^{-a}) I_k + R_k, \tag{7.4.79}$$

where transfer out of the infective population to a removed population, R_k, has been added. A new positive parameter, a, has been introduced, and it follows that

$$0 < e^{-a} < 1, \quad a > 0. \tag{7.4.80}$$

To proceed, biological meaning must be given to Equations (7.4.78) to (7.4.80). We interpret them as follows:

(A) The equation for the susceptible population includes a net birthrate, $B > 0$. Thus, S_{k+1} is equal to the new additions coming from births and the susceptibles who were not infected. This means that the exponential term, $\exp(-\beta I_k)$, plays the role of a probability for not becoming infective.

(B) The population of infectives at time $(k + 1)$ is equal to these susceptible members who became infected, added to the number of the infected population who did not transfer out to the removed population, R_k. As it appears in Equation (7.4.79), the factor, $\exp(-a)$, is the probability for an infected individual not to transfer to the removed population.

(C) The interpretation of Equation (7.4.80) is now obvious.

If Equations (7.4.78), (7.4.79), and (7.4.80) are added, then we obtain

$$P_{k+1} = P_k + B, \tag{7.4.81}$$

where

$$P_k = S_k + I_k + R_k, \tag{7.4.82}$$

and

$$P_k = P_o + Bk. \tag{7.4.83}$$

The total population increases linear in the discrete time.

To construct an appropriate corresponding set of differential equation, we introduce a new parameter, $h = \Delta t$, and rescale the other parameters as follows

$$B = h\bar{B}, \quad \beta = h\bar{\beta}, \quad a = h\bar{a}. \tag{7.4.84}$$

For "small" h, we have from Equation (7.4.78), the result

$$S_{k+1} = h\bar{B} + S_k\left[1 - h\bar{\beta}I_k + O\left(h^2\right)\right], \tag{7.4.85}$$

and this can be rewritten as

$$\frac{S_{k+1} - S_k}{h} = \bar{B} = \bar{\beta}S_k I_k + O\left(h^2\right). \tag{7.4.86}$$

Taking the limit, $h \to 0$, gives

$$\frac{dS}{dt} = \bar{B} - \bar{\beta}S\,I, \tag{7.4.87}$$

where $S = S(t)$ and $I = I(t)$.

If the same limiting process is done for Equations (7.4.79) and (7.4.80), then we find

$$\frac{dI}{dt} = \bar{\beta}\,S\,I - \bar{a}\,I \tag{7.4.88}$$

$$\frac{dR}{dt} = \bar{a}\,I, \tag{7.4.89}$$

with

$$\frac{dP}{dt} = \bar{B},\, P(t) = S(t) + I(t) + R(t). \tag{7.4.90}$$

We can also construct corresponding differential equations to a second modification of the Anderson-May discrete time model. For this case, the modified equations come from the
relations

$$S_k - \beta S_k I_k = S_k(1 - \beta I_k) \to \frac{S_k}{1 + \beta S_k}, \tag{7.4.91}$$

$$\beta S_k I_k \to \frac{\beta S_k I_k}{1 + \beta I_k}. \tag{7.4.92}$$

Therefore, we have

$$S_{k+1} = B + \left(\frac{1}{1 + \beta I_k}\right)S_k, \tag{7.4.93}$$

$$I_{k+1} = \left(\frac{\beta I_k}{1 + \beta I_k}\right)S_k + (e^{-a})\,I_k, \tag{7.4.94}$$

$$R_{k+1} = R_k + (1 - e^{-a})\,I_k, \tag{7.4.95}$$

where $a > 0$. It follows that

$$P_{k+1} = P_k + B, \quad P_k = S_k + I_k + R_k. \tag{7.4.96}$$

and

$$P_k = P_o + Bk. \tag{7.4.97}$$

The continuum limit yields the same set of differential equations as was derived for the first modified model.

7.5 PARTIAL DIFFERENTIAL EQUATION APPLICATIONS

7.5.1 Linear Advection-Diffusion Equation

This partial differential equation is

$$u_t + u_x = bu_{xx}, \quad u = u(x,t), \tag{7.5.1}$$

where $b \geq 0$ and we use the notation

$$u_t \equiv \frac{\partial u}{\partial t}, \quad u_x \equiv \frac{\partial u}{\partial x}. \tag{7.5.2}$$

Physically, the three terms have the following meanings :

u_t : gives the evolution of u in time;

u_x : provides the movement of the system relative to the x-axis;

u_{xx} : describes the diffusion of system.

We assume that

$$u(x,0) \geq 0 \Rightarrow \quad u(x,t) \geq 0, \quad t > 0. \tag{7.5.3}$$

Our notation is as follows

$$\begin{cases} x \to x_m = (\Delta x)m, \quad t \to t_k = (\Delta t)k, \\ u(x,t) \to u_m^k, \end{cases} \tag{7.5.4}$$

where

$$m = \{\text{integers}\}, \quad k = (0, 1, 2, \dots, \}. \tag{7.5.5}$$

Inspection of Equation (7.5.1) indicates that it has the following three sub-equations,

$$u_t + u_x = 0, \tag{7.5.6}$$

$$u_x = bu_{xx}, \tag{7.5.7}$$

$$u_t = bu_{xx}. \tag{7.5.8}$$

The first two sub-equations have exact finite difference schemes and they are

$$\frac{u_m^{k+1} - u_m^k}{\Delta t} + \frac{u_m^k - u_{m-1}^k}{\Delta x} = 0, \ \Delta x = \Delta t, \tag{7.5.9}$$

$$\frac{u_m - u_{m-1}}{\Delta x} = b \cdot \frac{u_{m+1} - 2u_m + u_{m-1}}{b\left[\exp\left(\frac{\Delta x}{b}\right) - 1\right]\Delta x}. \tag{7.5.10}$$

Observe that the discrete space derivative has the same structure in both schemes. Therefore, they can be combined to give the NSFD discretization

$$\frac{u_m^{k+1} - u_m^k}{\Delta t} + \frac{u_m^k - u_{m-1}^k}{\Delta x} = b \cdot \frac{u_{m+1}^k - 2u_n^k + u_{m-1}^k}{b\left[\exp\left(\frac{\Delta x}{b}\right) - 1\right]\Delta x}. \tag{7.5.11}$$

Solving for u_m^{k+1} gives the expression

$$u_m^{k+1} = \beta u_{m+1}^k + (1 - \alpha - 2\beta) u_m^k + (\alpha + \beta) u_{m-1}^k, \tag{7.5.12}$$

where

$$\alpha = \frac{\Delta t}{\Delta x}, \quad \beta = \frac{\alpha}{\exp\left(\frac{\Delta x}{b}\right) - 1}. \tag{7.5.13}$$

Note that while the scheme of Equation (7.5.9) requires $\Delta x = \Delta t$, this condition will not, in general, hold for the NSFD scheme listed in Equation (7.5.11). We wish to have the positivity condition hold, i.e.,

$$u_m^k \geq 0 \Rightarrow u_m^{k+1} \geq 0, \text{ all } m. \tag{7.5.14}$$

To enforce this requirement, we take

$$1 - \alpha - 2\beta \geq 0, \tag{7.5.15}$$

or

$$\Delta t \leq \left[\frac{\exp\left(\frac{\Delta x}{b}\right) - 1}{\exp\left(\frac{\Delta x}{b}\right) + 1} \right] \Delta x. \tag{7.5.16}$$

If we take the Lim $b \to 0$, then we obtain from Equation (7.5.16) the result

$$\Delta t \leq \Delta x, \tag{7.5.17}$$

and Equation (7.5.1) reduces to Equation (7.5.6). This implies that the equal sign should be used in Equation (7.5.16), i.e., the NSFD scheme given by Equation (7.5.11) is valid provided Δt and Δx satisfy the relation

$$\Delta t = \left[\frac{\exp\left(\frac{\Delta x}{b}\right) - 1}{\exp\left(\frac{\Delta x}{b}\right) + 1} \right] \Delta x. \tag{7.5.18}$$

Finally, observe that for $\Delta x / b$ small, i.e.,

$$\frac{\Delta x}{b} \ll 1, \tag{7.5.19}$$

then the step sizes relation, see Equation (7.5.18) becomes

$$\Delta t \simeq \frac{(\Delta x)^2}{2b}, \tag{7.5.20}$$

Also, for $\Delta x / b$ large, i.e.,

$$\frac{\Delta x}{b} \gg 1, \tag{7.5.21}$$

then

$$\Delta t \simeq \Delta x. \tag{7.5.22}$$

These results have the following interpretation:

(i) When, $\Delta x / b \ll 1$, the NSFD scheme acts like it is a diffusion equation.

(ii) For, $\Delta x / b \gg 1$, the NSFD scheme mimics as an advection equation.

7.5.2 Fisher Equation

The fisher partial differential equation

$$u_t = u_{xx} + \lambda u(1-u), \quad u = u(x, t), \quad \lambda > 0, \tag{7.5.23}$$

appears in many areas of the natural sciences. We now construct a NSFD discretization.

The two important sub-equations for our purposes are

$$u_t = \lambda u(1-u), \tag{7.5.24}$$

$$u_{xx} \neq \lambda u = 0. \tag{7.5.25}$$

They have, respectively, the following exact finite difference schemes

$$\frac{u^{k+1} - u^k}{\left(\frac{e^{\lambda \Delta t} - 1}{\lambda}\right)} = \lambda u^k \left(1 - u^{k+1}\right), \tag{7.5.26}$$

$$\frac{u_{m+1} - 2u_m + u_{m-1}}{\left(\frac{4}{\lambda}\right) \sin^2\left(\frac{\sqrt{\lambda}\Delta x}{2}\right)} + \lambda u_m = 0. \tag{7.5.27}$$

Combining these two expressions gives for the Fisher equation the discretization

$$\frac{u_m^{k+1} - u_m^k}{\left(\frac{e^{\lambda \Delta t} - 1}{\lambda}\right)} = \frac{u_{m+1}^k - 2u_m^k + u_{m-1}^k}{\left(\frac{4}{\lambda}\right) \sin^2\left(\frac{\sqrt{\lambda}\Delta x}{2}\right)} + \lambda u_m^k - \lambda \left(\frac{u_{m+1}^k + 2u_m^k + u_{m-1}^k}{4}\right) u_m^{k+1}. \tag{7.5.28}$$

Note that the λu^2 term has the discretization

$$\lambda u^2 = \lambda u u \to \lambda \left(\frac{u_{m+1}^k + 2u_m^k + u_{m-1}^k}{4}\right) u_m^{k+1}, \tag{7.5.29}$$

i.e., one "u" is replaced by an average over the discrete space variable at the time, t_k, while the second "u" is evaluated at the time, t_{k+1}. Thus, this way of discretizing "u" is non-local in both the discrete and time variables.

If we solve for u_m^{k+1}, the find

$$u_m^{k+1} = \left[\frac{1}{1 + (e^{\lambda \Delta t} - 1)}\right] \left\{ \left(\frac{\phi_1}{\phi_2}\right) \left(u_{m+1}^k + u_{m-1}^k\right) + \left[e^{\lambda \Delta t} - 2\left(\frac{\phi_1}{\phi_2}\right)\right] u_m^k \right\}, \tag{7.5.30}$$

where

$$\overline{u_m^k} = \frac{u_{m+1}^k + 2u_m^k + u_{m-1}^k}{4}, \tag{7.5.31}$$

$$\phi_1 = \phi_1(\Delta t) = \frac{e^{\lambda\Delta t} - 1}{\lambda}, \tag{7.5.32}$$

$$\phi_2 = \phi_2(\Delta x) = \left(\frac{4}{\lambda}\right)\sin^2\left(\frac{\sqrt{\lambda}\Delta x}{2}\right). \tag{7.5.33}$$

Now, the positivity condition will hold, if the coefficient of the u_m^k term is non-negative. Setting it to zero gives

$$e^{\lambda\Delta t} = 2\frac{\phi_1(\Delta t)}{\phi_2(\Delta x)}, \tag{7.5.34}$$

and then solving for Δt produces the relation

$$\Delta t = \left(\frac{-1}{\lambda}\right)\text{Ln}\left[\cos\left(\sqrt{\lambda}\Delta x\right)\right]. \tag{7.5.35}$$

Inspection of this equation gives a restriction on the size of Δx, namely,

$$\Delta x < \left(\frac{\pi}{2}\right)\left(\frac{1}{\sqrt{\lambda}}\right), \tag{7.5.36}$$

and for small Δx and Δt, the additional relation

$$\Delta t \approx \frac{(\Delta x)^2}{2}. \tag{7.5.37}$$

7.5.3 Combustion Model

The partial differential equation

$$u_t = u_{xx} + u^2(1 - u), \tag{7.5.38}$$

provides an elementary model of combustion in one space dimension. It may also be thought of as a Fisher type equation with a different reaction term, i.e., $u^2(1 - u)$ rather than $u(1 - u)$.

The physical meaningful solutions are those having the property

$$0 \le u(x, 0) \le 1 \;\Rightarrow\; 0 \le u(x, t) \le 1,\; t > 0. \tag{7.5.39}$$

A (possible) NSFD scheme for the combustion equation is

$$\begin{aligned}\frac{u_m^{k+1} - u_m^k}{\Delta t} &= \frac{u_{m+1}^k - 2u_m^k + u_{m-1}^k}{(\Delta x)^2} \\ &+ \left[\left(u_{m+1}^k\right)^2 + \left(u_{m-1}^k\right)^2\right] - \left[\frac{u_{m+1}^k + u_{m-1}^k}{2}\right] u_m^{k+1} \\ &- \left[\frac{\left(u_{m+1}^k\right)^2 + \left(u_{m-1}^k\right)^2}{2}\right] u_m^{k+1}.\end{aligned} \tag{7.5.40}$$

This scheme was constructed by making use of the following facts and restrictions:

(i) Equation (7.5.38) is invariant under the transformation

$$x \to x. \tag{7.5.41}$$

In terms of the discrete variable, x_m, this transformation is equivalent to having the discretization invariant under the index-m interchange

$$(m+1) \leftrightarrow (m-1) \tag{7.5.42}$$

Note, all the discretized terms must have this property.

(ii) The second-order space derivative is modelled by a central difference scheme

$$u_{xx} \to \frac{u_{m+1}^k - 2u_m^k + u_{m-1}^k}{(\Delta x)^2}, \tag{7.5.43}$$

(iii) The u^2 term has the discretization

$$u^2 = 2u^2 - u^2 \to \left(u_{m+1}^k\right)^2 + \left(u_{m-1}^k\right)^2 - \left(\frac{u_{m+1}^k + u_{m-1}^k}{2}\right) u_m^{k+1}. \tag{7.5.44}$$

(iv) For u^3, we use

$$u^3 \to \left[\frac{(u_{m+1}^k)2 + \left(u_{m-1}^k\right)^2}{2}\right] u_m^{k+1}. \tag{7.5.45}$$

Putting all these features together produces the NSFD scheme represented in Equation (7.5.40). Since in this equation, u_m^{k+1}, appears linearly, we can solve for it and obtain the expression

$$u_m^{k+1} = \frac{R\left(u_{m+1}^k + u_{m-1}^k\right) + (\Delta t)\left[\left(u_{m+1}^k\right)^2 + \left(u_{m-1}^k\right)^2\right] + (1-2R)u_m^k}{1 + \left(\frac{\Delta t}{2}\right)\left[u_{m+1}^k + \left(u_{m+1}^k\right)^2 + u_{m-1}^k + \left(u_{m-1}^k\right)^2\right]}, \tag{7.5.46}$$

where

$$R = \frac{\Delta t}{(\Delta x)^2}. \tag{7.5.47}$$

Note that u_m^{k+1} is certainly non-negative if

$$1 - 2R \geq 0, \tag{7.5.48}$$

If the equal sign is selected, then the NSFD scheme becomes a little simplier and we have

$$\Delta t = \frac{(\Delta x)^2}{2}, \quad R = \frac{1}{2}. \tag{7.5.49}$$

For this case, it can be shown that the following result is true for all m and fixed k:

$$0 \leq u_m^k \leq 1 \Rightarrow 0 \leq u_m^{k+1} \leq 1. \tag{7.5.50}$$

The proof of this result makes use of the following fact

$$0 \leq w \leq 1 \Rightarrow w^2 \leq \frac{w + w^2}{2}. \tag{7.5.51}$$

7.6 DISCUSSION

In its present stage of development, the NSFD methodology has a large component of creativity attached to its applications. For example, given a function, $x(t)$, how should it be non-locally discretized? Possibilities include (for a first-order differential equation)

$$x(t) \to \begin{cases} 2x_k - x_{k+1}, \\ \dfrac{x_k + x_{k+1}}{2}. \end{cases} \tag{7.6.1}$$

Another issue is related to the solution of denominator functions for the discrete derivatives, i.e., given a system of n coupled first-order differential equations

$$\frac{dx}{dt} = F(x), \tag{7.6.2}$$

$$x(t) = \begin{pmatrix} x_1(t) \\ x_2(t) \\ \vdots \\ x_N(t) \end{pmatrix}, F(x) = \begin{pmatrix} f_1(x) \\ f_2(x) \\ \vdots \\ f_N(t) \end{pmatrix}, \tag{7.6.3}$$

we would like to have the discrete derivatives take the form

$$\frac{dx_i(t)}{dt} \to \frac{x_{k+1}(i) - x_k(i)}{\phi(h)}, \tag{7.6.4}$$

$i = (1, 2, \dots N)$, where there is only one denominator function, $\phi(h)$. How should $\phi(h)$ be selected?

All the differential equations discussed in Sections 7.4 and 7.5 have solutions that must satisfy a positivity condition. This is a consequence of the fact that these equations model physical entities such as particle numbers, population densities, absolute temperatures, etc., that by their definition and/or meaning can not be negative. But what does one do for cases when this is not true? For instance, the van der Pol oscillator is modelled by

$$\frac{d^2x}{dt^2} + \Omega^2 x = \epsilon\left(1 - x^2\right)\frac{dx}{dt}, \tag{7.6.5}$$

or, in system form

$$\frac{dx}{dt} = y, \quad \frac{dy}{dt} = -\Omega^2 x + \epsilon\left(1 - x^2\right)y. \tag{7.6.6}$$

Two issues immediately arise:

(a) If the system equations are used, then what are the discrete NSFD analogues for the two derivatives dx/dt and dy/dt?

(b) Which provides a "better" discretizations, the single, second-order differential equation or the coupled pair of equations?

An interesting topic, that has not been adequately studied, is the construction and investigation of, what we will call, algebraically formulated finite difference schemes. As a simple illustration, consider the equation

$$\frac{dx}{dt} = -x^{\frac{1}{3}}, \quad x(0) = x_0 > 0, \tag{7.6.7}$$

with the discretization

$$\frac{x_{k+1} - x_k}{h} = -(x_{k+1})^{\frac{1}{3}}. \tag{7.6.8}$$

If y_k is defined to be

$$y_k = (x_k)^{\frac{1}{3}}, \tag{7.6.9}$$

then

$$(y_{k+1})^3 + hy_{k+1} - x_k = 0. \tag{7.6.10}$$

This is a cubic equation in y_{k+1}, with one real root positive. Therefore, there exists a function

$$G(h, x_k)^{\text{such that}} y_{k+1} = G(h, x_k). \tag{7.6.11}$$

Therefore, the finite difference discretization of Equation (7.6.7) is

$$x_{k+1} = [G(h, x_k)]^3, \quad x_0 > 0. \tag{7.6.12}$$

Finally, it must be stated and clearly understood that for a given system of differential equations, the NSFD representations are not unique. A way to see this is to note that a particular differential equation may have a number of sub-equations and these sub-equations may be combined in several inequivalent ways to formulate different NSFD schemes.

PROBLEMS

Section 7.1

1) Consider the harmonic oscillator differential equation

$$\frac{d^2x}{dt} + x = 0.$$

Three discrete finite different models are

$$\frac{x_{k+1} - 2x_k + x_{k-1}}{h^2} + x_{k-1} = 0,$$

$$\frac{x_{k+1} - 2x_k + x_{k-1}}{h^2} + x_k = 0,$$

$$\frac{x_{k+1} - 2x_k + x_{k-1}}{h^2} + x_{k+1} = 0.$$

Analysis the stability properties of these three schemes and determine which, if any, can be used to model the harmonic oscillator differential equation.

Section 7.2

2) Construct an exact finite difference scheme for the Michalis-Menton equation

$$\frac{dx}{dt} = -\frac{ax}{1+bx}.$$

(This is a difficult problem.)

3) Calculate the solution to the equation

$$\frac{dx}{dt} = x + \frac{1}{x}, \quad x(t_0) > 0,$$

and from this result construct the corresponding exact finite difference scheme.

4) Every ordinary differential equation has as least one finite difference scheme. Show, by example, that there exist difference equations that have no corresponding differential equation model.

Section 7.3

5) Consider the decay equation

$$\frac{dx}{dt} = -\lambda x.$$

Show that the following two schemes are exact finite difference representations for this
equation

$$\frac{x_{k+1} - x_k}{\left(\frac{1-e^{-\lambda h}}{\lambda}\right)} = -\lambda x_k,$$

$$\frac{x_{k+1} - x_k}{\left(\frac{e^{\lambda h}-1}{\lambda}\right)} = -\lambda x_{k+1}.$$

What is the relationship between these schemes.

Section 7.4

6) Construct a NSFD scheme for

$$\frac{d^2x}{dt^2} + x^{\frac{1}{3}} = 0.$$

Section 7.5

7) Obtain an exact solution to the nonlinear partial differential equation

$$\frac{\partial u}{\partial t} + \frac{\partial u}{\partial x} = u(1-u)$$

and use it to construct an exact finite difference scheme. Use the NSFD scheme methodology to construct a scheme for this equations. In what significant ways do these schemes differ?

NOTES AND REFERENCES

Section 7.1, 7.2, and 7.3: The following books give general discussions on standard discretizations of differential equations

(1) J.C. Butcher, **The Numerical Analysis of Ordinary Differential Equations** (Wiley-Interscience, Chichester, 1987).

(2) F.B. Hildebrand, **Finite-Difference Equations and Simulations** (Prentice-Hall; Englewood Cliffs, NJ; 1968).

(3) M.K. Jain, **Numerical Solutions of Differential Equations**, 2nd Edition (Wiley, New York, 1984).

(4) G.D. Smith, **Numerical Solutions of Partial Differential Equations** (Clarendon Press,
Oxford, 1978).

(5) J.C. Strikwerda, **Finite Difference Schemes and Partial Differential Equations** (Clarendon Press Oxford, 1978).

The nonstandard methodology and its applications are given in the following books and review article

(6) R.E. Mickens (editor), **Advances in the Applications of Nonstandard Finite Difference Schemes** (World Scientific, Singapore, 2005).

(7) R.E. Mickens, **Nonstandard Finite Difference Schemes: Methodology and Applications** (World Scientific, London, 2021).

(8) K.C. Patidar, Nonstandard finite difference methods: Recent trends and further developments, **Journal of Difference Equations and Applications**, Vol. 22 (2016), 817-849.

A good general introduction to finite difference equations is the book

(9) R.E. Mickens, **Difference Equations: Theory and Applications** (Van Nostrand Reinhold, New York, 1990).

Section 7.4 and 7.5: The examples discussed in these sections come directly from Mickens (1990, 2005, 2021).
Section 7.6: For many applications, the denominator functions can be explicitly calculated. See

(10) R.E. Mickens, Calculation of denominator functions for NSFD schemes for differential equations satisfying a positivity condition, **Numerical Methods for Partial Differential Equations**, Vol. 32 (2007), 672-691.

(11) R.E. Mickens, Numerical integration of population models satisfying conservation laws: NSFD methods, **Journal of Biological Dynamics**, Vol. 1 (2007), 427-436.

CHAPTER 8

SIR Models for Disease Spread

8.1 INTRODUCTION

The major purpose of this chapter is to construct a mathematical model for communicable diseases such as measles, influenza, and COVID-19. The simplest such model is has a SIR representation, i.e., the total population is taken to be constant during the disease outbreak and is divided into three sub-populations:

S(t): susceptible individuals, i.e., those members of the population who can become infected, but are not;

I(t): infected individuals who upon contact with susceptible individuals can infect the susceptible persons;

R(t): recovered and/or removed individuals who have either recovered from the disease or have been removed themselves by dying, etc.

The function of the mathematical model is to predict the evolution of the disease as transitions are made from one population class to another, i.e.,

$$S(t) \to I(t) \to R(t). \tag{8.1.1}$$

Mathematical models are important because they allow us to understand the current state of a system and predict its future states. However, different models may provide different outcomes and this flexibility can also give insights into fundamental features of the system. It must be understood that the models are not the actual system. They are abstract mathematical representations of only some of its important features, the ones of relevance for our needs.

In more detail, the purposes of this chapter are several:

(i) Show the construction of a SIR model where the total population is taken to be constant. This model will consist of three coupled ordinary differential equations and is the simplest of the SIR class of models.

DOI: 10.1201/9781003178972-8

(ii) Provide appropriate interpretations of the parameters appearing in the model and their connection to epidemiological data.

(iii) Derive a number of the significant mathematical properties of the solutions to the model, in spite of not being able to explicitly obtain these solutions.

(iv) Give a direct, simple geometrical explanation of the effects of "stay-at-home-orders" and what occurs when they are relaxed.

(v) Finally, show that a simple SIR mathematical model allows detailed qualitative predictions to be made for the general features of diseases such as influenza and COVID-19.

This chapter has the following organization: Section 8.2 presents the details of what the three variables, $(S(t), I(t), R(t))$, mean, and the general mathematical structures denoting how these three variables are related. In Section 8.3, we examine the standard SIR model and calculate and discuss the properties of its solutions. Section 8.4 gives a geometric interpretation of what occurs when stay-at-home orders are imposed on the population. Using different population interaction terms, we construct in Section 8.5 a model that can be exactly solved in terms of elementary functions.

8.2 SIR METHODOLOGY

The construction of SIR mathematical models is based on the following concepts and assumptions:

(a) The total population is composed of three sub-populations, susceptibles, $S(t)$; infected, $I(t)$; and recovered, $R(t)$. Susceptibles are uninfected and susceptible to the disease; the infected population is, by definition, infected and can in turn infect susceptibles; and the recovered individuals have recovered from the disease and/or died and now are immune to reinfection.

(b) The total population is assumed to be constant, i.e.,

$$N = S(t) + I(t) + R(t) = \text{constant} \tag{8.2.1}$$

This constraint implies that over the time interval for which the model is relevant, the birth and death rates are equal.

(c) It is assumed that the three sub-populations are homogeneous mixed. This means that all individuals in the total population have the same probability of coming into contact with each other an interacting.

(d) It is assumed that the disease can be transmitted between two individuals regardless of their age.

The explicit construction of a SIR model begins when a framework is formulated for how the three populations transfer from one sub-population to another. Based on the schema stated in Equation (8.1.1), the general mathematical structure of an SIR model takes the following form

$$\frac{dS(t)}{dt} = -T_1(S \to I), \tag{8.2.2}$$

$$\frac{dI(t)}{dt} = T_1(S \to I) - T_2(I \to R), \tag{8.2.3}$$

$$\frac{dR(t)}{dt} = T_2(I \to R), \tag{8.2.4}$$

where

$T_1(S \to I)$: transition rate from the S population to the I population;

$T_2(I \to R)$: transition rate from the I population to the R population.

Note that adding Equations (8.2.2), (8.2.3), and (8.2.4) gives

$$\frac{d}{dt}[S(t) + I(t) + R(t)] = 0, \tag{8.2.5}$$

or

$$\begin{aligned} N(t) &\equiv S(t) + I(t) + R(t) \\ &= S_0 + I_0 + R_0 \\ &= \text{constant}, \end{aligned} \tag{8.2.6}$$

where $N(t)$ is the total population and

$$S(0) = S_0, I(0) = I_0, R(0) = R_0. \tag{8.2.7}$$

With the placement of the negative signs in the rate equation, it follows that

$$S \geq 0,\ I \geq 0 \Rightarrow \begin{cases} T_1(S \to I) \geq 0, \\ T_2(I - R) \geq 0. \end{cases} \tag{8.2.8}$$

In somewhat more detail, we can provide another level of discussion for Equations (8.2.2), (8.2.3), and (8.2.4). The dS/dt equation implies that each S that gets infected gets transferred into the I-population. Consequently, the S-population decreases and the I-population increases. From the dI/dt equation, we see that the first term on the right-side corresponds to additions to the I-population coming from the newly infected members of the S-population. The second term represents those members of the I-population that have recovered from the disease and are now transferred to the R-population.

Careful reflection on the dynamics of a general SIR system allows the following conclusions to be drawn for the transition functions, T_1 and T_2:

(i) Both $T_1(S \to I)$ and $T_2(I \to R)$ are functions of S and I, and I, i.e.,

$$T_1(S \to I) = T_1(S, I), \tag{8.2.9}$$

$$T_2(I \to R) = T_2(I). \tag{8.2.10}$$

(ii) $T_1(S,I)$ has the properties

$$\begin{cases} T_1(S,I) > 0, \text{ for } S > 0,\ I > 0; \\ T_1(0,I) = T_1(S,0) = 0; \\ T_1(S,I) \text{ is a monotonic increasing function of } S \text{ and } I. \end{cases} \tag{8.2.11}$$

(iii) $T_2(I)$ has the properties

$$\begin{cases} T_2(I) > 0, \text{ for } I > 0; \\ T_2(0) = 0; \\ T_2(I) \text{ is a monotonic increasing function of } I. \end{cases} \tag{8.2.12}$$

It should be note, with some care, that there are many functions satisfying the above conditions imposed on $T_1(S,I)$ and $T_2(I)$. Some examples appearing in the research literature are $T_1(S,I)$

$$\beta_1 S I, \quad \beta_2 \sqrt{S}\sqrt{I}, \tag{8.2.13a,b}$$

$$\frac{\beta_3 S I}{1+aI}, \quad \left(1 - e^{-\beta I}\right) S. \tag{8.2.13c,d}$$

$T_2(I)$

$$\gamma_1 I, \quad \gamma_2 \sqrt{I}, \tag{8.2.14a,b}$$

$$1 - e^{-\beta I}, \tag{8.2.14c}$$

where all the parameters, $(\beta, \beta_1, \beta_2, a)$ are non-negative.

8.3 STANDARD SIR MODEL

The standard SIR model uses the following forms for the transition functions

$$T_1(S,I) = \beta S\left(\frac{I}{N}\right), \quad T_2(I) = \gamma I, \tag{8.3.1}$$

which gives for the sub-populations rate equations the three expressions

$$\frac{dS}{dt} = -\beta S\left(\frac{I}{N}\right), \tag{8.3.2}$$

$$\frac{dI}{dt} = \beta S\left(\frac{I}{N}\right) - \gamma I, \tag{8.3.3}$$

$$\frac{dR}{dt} = \gamma I, \tag{8.3.4}$$

where $N = S + I + R =$ constant, and (β, γ) are constant parameters which can be related to several important aspects of the evolution of the disease in the full population.

With regard to the last point, observe that the left-sides of Equations (8.3.2), (8.3.3) and (8.3.4) all have the units of population number over time. This means that each of the individual terms on the right-sides must have the same units. Consequently, in this particular SIR model both β and γ have the units of inverse time. From this fact, these two parameters can be given the following interpretations:

$t_c \equiv \frac{1}{\beta}$ = average time between contacts of the S and I sub-populations;

$t_r \equiv \frac{1}{\gamma}$ = average time a member of the infected population stays infected and then gets transferred to the removed population.

Since the rate equations for S and I do not involve R, our analysis need only involve the differential equations for S and I, i.e.,

$$\frac{dS}{dt} = -\beta S\left(\frac{I}{N}\right), \quad \frac{dI}{dt} = \beta S\left(\frac{I}{N}\right) - \gamma I. \tag{8.3.5}$$

The second of these equation may be rewritten as

$$\frac{dI}{dt} = \beta\left(\frac{I}{N}\right)(S - S^*), \tag{8.3.6}$$

where

$$\begin{aligned} S^* &= \left(\frac{\gamma}{\beta}\right)N \\ &= \left(\frac{t_c}{t_r}\right)N. \end{aligned} \tag{8.3.7}$$

If $S(t)$ and $I(t)$ are known, then $R(t) = N - S(t) - I(t)$. Also, please keep in mind that

$$\frac{dI}{dt}: \begin{cases} > 0 \Rightarrow I(t) \text{ is increasing;} \\ = 0 \Rightarrow I(t) \text{ is stationary;} \\ < 0 \Rightarrow I(t) \text{ is decreasing.} \end{cases}$$

Further, all three sub-populations are non-negative, i.e.,

$$0 \leq S(t) \leq N, \ 0 \leq I(t) \leq N, \ 0 \leq R(t) \leq N. \tag{8.3.8}$$

With this information, the following conclusions can be reached from an inspection of Equation (8.3.6):

(A) Let at time, $t = 0$, $S_0 > 0$ and $I_0 > 0$, but with $I_0 << S_0$. If $S_0 > S^*$, then $I(t)$, for $t > 0$, will initially increase, but, $S(t)$ will monotonically decrease for all $t > 0$. At some future time, $S(t)$ will fall to a value below S^* and when $S = S^*$, $I(t)$ will begin to decrease and go to zero. This situation is depicted in Figure 8.1(a). For this case, we have a classical epidemic.

(B) For the same initial conditions as in (A), let $S_0^1 < S^*$; then $I(t)$ decreases to zero as indicated in Figure 8.1(b). No epidemic occurs.

The situations depicted in (A) and (B) give rise to the famous threshold theorem of epidemiology:

> The placement of a single infective in a susceptible population will only initiate an epidemic if the number of susceptible is larger than a certain threshold value, which in our case is S^*.

Another way of stating this result is to note this statement is equivalent to the condition that the rate at which susceptibles become infectives must be larger than the rate which infectives are removed from the population in order to have an epidemic take place.

Define r_0 as follows

$$r_0 \equiv \left(\frac{\beta}{\gamma}\right)\left(\frac{S_0}{N}\right) = \frac{S_0}{S^*}. \tag{8.3.9}$$

This parameter is called the "basic reproduction number" and plays a fundamental role in SIR based epidemiology. It is generally **interpreted** as the average number of secondary infections caused by introducing a single infective individual into a susceptible population. It is easily seen that if $r_0 > 1$, then an epidemic will occur, while for $r_0 < 1$, the infective population decreases from the beginning and no epidemic occurs.

The curve of I versus S provides additional insights on the dynamics of the SIR model. If this curve is denoted by

$$I = I(S), \tag{8.3.10}$$

then

$$\frac{dI}{dt} = \frac{dI}{dS}\frac{dS}{dt}, \tag{8.3.11}$$

so

$$\begin{aligned} \frac{dI}{dS} &= \frac{dI/dt}{dS/dt} \\ &= -1 + \frac{S^*}{S}. \end{aligned} \tag{8.3.12}$$

For the last step, see the relations given in Equation (8.3.5). Therefore,

$$dI = -dS + S^* \frac{dS}{S}, \tag{8.3.13}$$

and integrating, from $t = 0$ to t, gives

$$\int_{I_0}^{I} d\bar{I} = -\int_{S_0}^{S} d\bar{S} + S^* \int_{S_0}^{S} \frac{d\bar{S}}{\bar{S}}, \tag{8.3.14}$$

with the result that

$$I(t) - I_0 = -(S(t) - S_0) + S^* \operatorname{Ln}\left(\frac{S(t)}{S_0}\right), \tag{8.3.15}$$

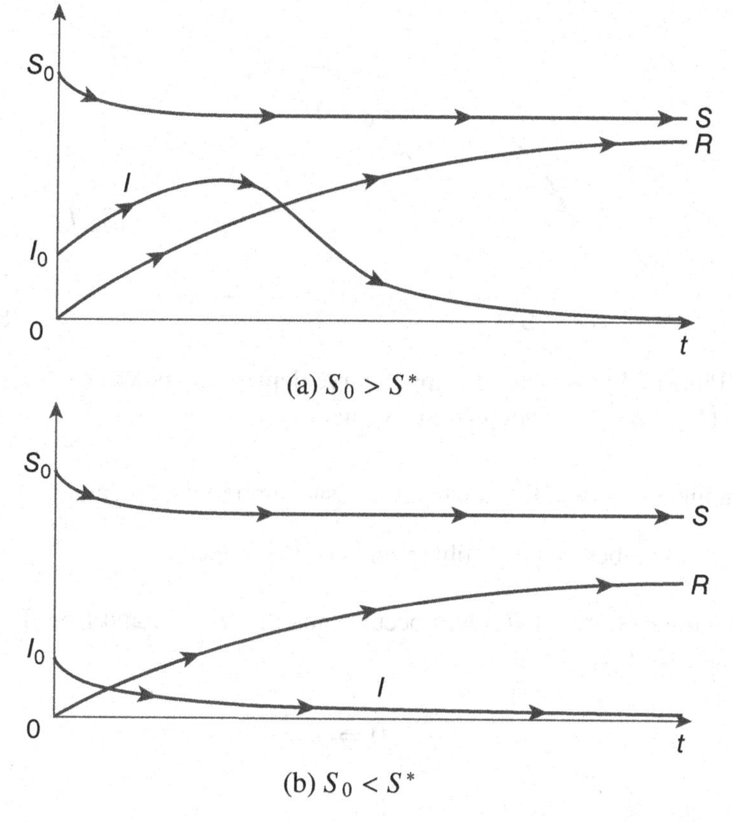

(a) $S_0 > S^*$

(b) $S_0 < S^*$

Figure 8.1: General time dependence of $S(t), I(t)$, and $R(t)$. (*a*) For $S_0 > S^*, r_0 > 1$. (*b*) For $S_0 < S^*, r_0 < 1$.

which upon rearrangement is the expression

$$I(t) + S(t) = I_0 + S_0 + S^* \operatorname{Ln}\left(\frac{S(t)}{S_0}\right). \tag{8.3.16}$$

The initial conditions are taken to be

$$S_0 = S(0) > 0, \ I_0 = I(0) > 0, R_0 = R(0) = 0. \tag{8.3.17}$$

Figure 8.2 is a plot of the I versus S curve for two situations:

(a) $S_0 > S^*$, corresponding to an epidemic.

(b) $S_0 < S^*$, for which no epidemic occurs.

These two cases match, respectively, the situation for $r_0 > 1$ and $r_0 < 1$.

Note that in terms of the time behaviour, the "motion" along these curves go from the right to the left sides of the graph and are correlated with the plots presented in Figure 8.1.

For the remainder of this section, only the case of an epidemic will be discussed. Of particular interest are the values of I_{max}, S_∞, and I_{total}, defined as

I_{max} = the maximum value of the number of infected persons during the course of the epidemic;

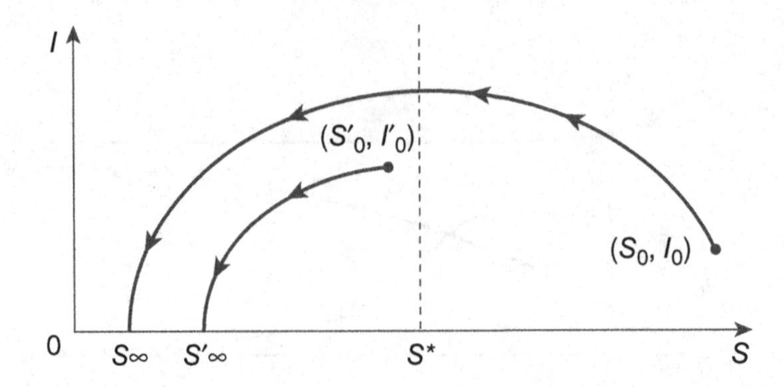

Figure 8.2: Plots of I versus S. The upper curve depicts an epidemic $(S_0 > S^*)$, while the lower curve $\left(S_0^* < S^*\right)$ does not lead to an epidemic.

S_∞ = the number of susceptibles who do not succumb to the epidemic;

I_{total} = the total number of susceptibles who become infected.

Now the maximum value of $I(t)$ will occur when $dI/dt = 0$ and from Equation (8.3.6), this takes place for $S = S^*$, i.e.,

$$\frac{dI}{dt} = 0 \Rightarrow S = S^*. \tag{8.3.18}$$

Thus, from Equation (8.3.16), it follows that

$$I_{\max} + S^* = I_0 + S_0 + S^* \text{Ln}\left(\frac{S^*}{S_0}\right), \tag{8.3.19}$$

and

$$I_{\max} = I_0 + (S_0 - S^*) + S^* \text{Ln}\left(\frac{S^*}{S_0}\right). \tag{8.3.20}$$

Likewise, at $t = \infty$, $I(\infty) = 0$, and

$$S_\infty = (I_0 + S_0) + S^* \text{Ln}\left(\frac{S_\infty}{S_0}\right). \tag{8.3.21}$$

While this is not one of the standard elementary equations, it can be solved by the introduction of the so-called Lambert-W function. Thus, S_∞ can be obtained, since we know I_0 and S_0, and can calculate S^* from β and γ.

Finally, the total number of persons who became infective is equal to the initial number of infectives introduced into the susceptive population, added to $S_0 - S_\infty$, i.e.,

$$I_{\text{total}} = I_0 + S_0 - S_0. \tag{8.3.22}$$

It is generally the case that at the initiation of an epidemic

$$I_0 << S^* < S_0, \tag{8.3.23}$$

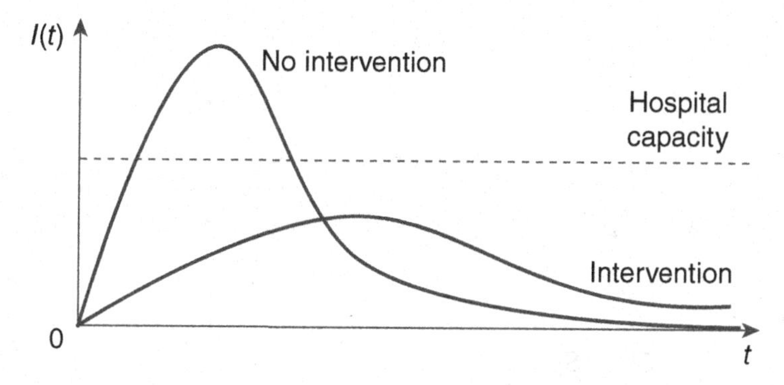

Figure 8.3: Flattening the curve: plot of $I(t)$ versus t for no intervention and for the outcome of good interventions.

and the above three expressions can be written, to a high degree of accuracy as

$$I_{\max} \simeq (S_0 - S^*) + S^* \mathrm{Ln}\left(\frac{S^*}{S_0}\right), \tag{8.3.24}$$

$$S_\infty \simeq S_0 + S^* \mathrm{Ln}\left(\frac{S_\infty}{S_0}\right), \tag{8.3.25}$$

$$I_{\text{total}} \simeq S_0 - S^*, \tag{8.3.26}$$

In summary, given the elementary SIR model where the parameters are assumed to be known, given values of S_0 and I_0, all of the general qualitative features of the epidemic can be determined, along with values for $I_{\max}, S_\infty$, and I_{total}. These results are important since for this model no explicit, known solutions exist in terms of a finite combination of the elementary functions.

8.4 FLATTENING THE CURVE

The curve that is being talked about is the plot of $I(t)$ versus t, i.e., the number of infectives as a function of time. So, how does this issue arise and what are its consequences?

Hospitals generally have a maximum capacity for treating acute illnesses in terms of the number of beds and available care teams. Further, some hospitals and emergency facilities may already be operating close to their maximum capacity under normal circumstances.

If the sharply peaked curve, see Figure 8.3, could be changed into the broader, flatter lower curve, lying below the full capacity curve of the hospital, then this would help hospitals provide better care for their patients and allow some relief to their emergency care staff. Thus, what we wish to achieve is a lowering of the total number of cases who are in the hospital at any given time to a number smaller than the maximum capacity and spread it out over a lorger period of time. This lowering and spreading can be done, in the absence of an actual effective vaccine through the use of physical distancing, stay-at-home orders, the wearing of face masks, and a number of other measures. All of these actions may be characterized as non-pharmaceutical interventions.

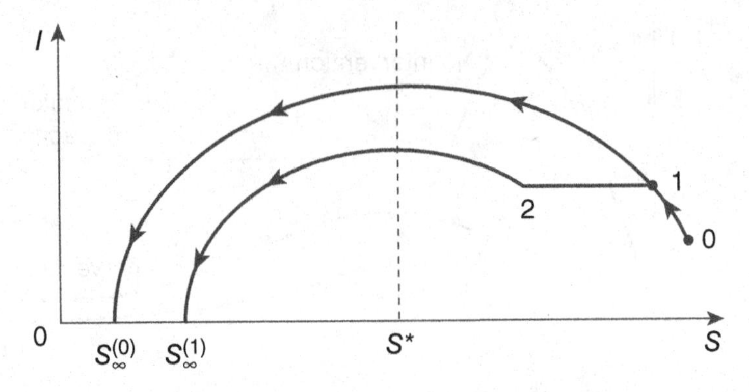

Figure 8.4: Consequences of the initiation of a stay-at-home order (beginning at 1; follow the heavy lines).

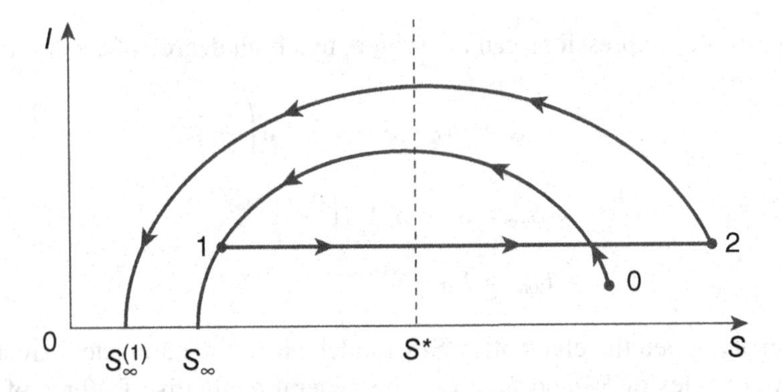

Figure 8.5: One possibility of halting a stay-at-home order (beginning at 1).

Let us now model the consequences of a particular non-pharmaceutical intervention on the outcomes of the SIR model. This can be done with graphic techniques as we now demonstrate.

The sole purpose of the intervention is to "flatten the curve" and for ease of interpretation and explanation, we consider a stay-at-home" order. For the immediate discussion, all comments and observations will be based on the curves in Figure 8.4.

The initiation of the epidemic is at point 0, where $P_0 = (S_0, I_0)$. The system then evolves to point $P_1 = (S_1, I_1)$ where

$$S_1 < S_0, \ I_1 > I_0. \tag{8.4.1}$$

At P_1, the stay-at-home order is made and we assume that some fraction of the susceptible population obeys this edict. (The exact value is not important; it just needs to remain constant for the arguments to be valid.) The system now goes from P_1 to $P_2 = (S_2, I_1)$, i.e., S decreases from S_1 to S_2, but the number of infectives is assumed to remain constant at the value I_1.

At P_2, the SIR system evolves in the usual manner, with $S(t)$ steadily decreasing to the value $S_\infty^{(1)}$ and $I(t)$, at first increasing to a peak and then decreasing to zero. As a function of t, $I(t)$ has the behavior shown in Figure 8.1(a).

A detailed examination of Figure 8.4 allows the following conclusions to be made:

(i) The evolution of the system with the stay-at-home order gives a smaller value for the maximum number of infectives in comparison to the situation of having no such order.

(ii) Since

$$S_{\infty}^{(1)} > S_{\infty}^{(0)}, \tag{8.4.2}$$

the total number of infections during the epidemic is reduced with the stay-at-home order.

(iii) Since the graphs in Figure 8.4 do not contain temporal information, we can not from just this information show that the peak in infectives occurs later for the case where a stay-at-home is in place; but, this is in fact what happens.

The trajectories in Figure 8.5 illustrate what may occur when a stay-at-home order is released. Assume that the path connecting P_0 and P_1 corresponds to a stay-at-home situation. Let the order be rescinded at $P_1 = (S_1, I_1)$. One possibility is that the system goes to point $P_2 = (S_2, I_1)$, reflecting an increase in the susceptible population due to susceptibles who stayed in their homes and, as a consequence, could not become infective. The system now evolves along the upper path with the number of infectives first increasing and then decreasing to zero. In other words, the release of new susceptibles causes a second, larger peak to occur in the number of infectives. This can be summarized in the following manner:

(1) The second peak in the infective curve is larger (in this scenario) after the stay-at-home order is dropped.

(2) After the epidemic has reached its course, the total infective population is also larger than would be the situation if the stay-at-home order was kept in place.

8.5 SOLVEABLE SIR MODEL

We now demonstrate that an SIR model can be constructed such that it can be explicitly solved in terms of elementary functions. This model consists of the following three, coupled differential equations

$$\frac{dS}{dt} = -\beta\sqrt{S}\sqrt{I}, \tag{8.5.1}$$

$$\frac{dI}{dt} = \beta\sqrt{S}\sqrt{I} - \gamma\sqrt{I}, \tag{8.5.2}$$

$$\frac{dR}{dt} = \gamma\sqrt{I}. \tag{8.5.3}$$

This model replaces the usual SIR model variables by their square roots, i.e.,

$$\beta S I \to \beta\sqrt{S}\sqrt{I}, \gamma I \to \gamma\sqrt{I}. \tag{8.5.4}$$

Immediately, a difference in the two models comes to light. If $S(t) = 0$, but $I_0 > 0$, then the standard model reduces to

$$\frac{dI}{dt} = -\gamma I, \quad I(0) = I_0, \tag{8.5.5}$$

which has the solution

$$I(t) = I_0\, e^{-\gamma t}. \tag{8.5.6}$$

However, in the new model, we have

$$\frac{dI}{dt} = -\gamma\sqrt{I},\, I(0) = I_0, \tag{8.5.7}$$

with solution

$$I(t) = \begin{cases} \left[\sqrt{I_0} - \dfrac{\gamma t}{2}\right]^2, & 0 < t \le t^*, \\ 0, \quad t > t^*, \end{cases} \tag{8.5.8}$$

where

$$t^* = \frac{2\sqrt{I_0}}{\gamma}. \tag{8.5.9}$$

Observe that different long time behaviors are predicted by the two models. The standard model gives $I(t) > 0$, for any $t > 0$, while the new model says that $I(t)$ decreases monotonic from $I = I_0$ to $I = 0$, in a finite time, t^*; for $t > t^*$, $I(t)$ is zero.

8.5.1 Conditions for an Epidemic

Equation (8.5.2) can be rewritten to the form

$$\frac{dI}{dt} = \beta\sqrt{S^*}\,\sqrt{I}\left(\sqrt{\frac{S}{S^*}} - 1\right), \tag{8.5.10}$$

where

$$S^* = \left(\frac{\gamma}{\beta}\right)^2. \tag{8.5.11}$$

Suppose $I(0) = I_0 > 0$ and $S(0) = S_0 > 0$, then

$$\left.\frac{dS}{dt}\right|_{t=0} = -\beta\sqrt{S_0}\,\sqrt{I_0}, \tag{8.5.12}$$

$$\left.\frac{dI}{dt}\right|_{t=0} = \beta\sqrt{S^*}\,\sqrt{I_0}\left(\sqrt{\frac{S_0}{S^*}} - 1\right). \tag{8.5.13}$$

From Equations (8.5.1) and (8.5.12), we conclude that $S(t)$ decreases from S_0, at $t = 0$, to some value S_∞ as $t \to \infty$. From Equation (8.5.13), we learn that if $S_0 < S^*$, then $dI/dt < 0$, for all $t > 0$, and consequently $I(t)$ decreases to zero, i.e.,

$$S_0 < S^* \Rightarrow \frac{dI}{dt} < 0 \Rightarrow \lim_{t\to\infty} I(t) = 0. \tag{8.5.14}$$

This means that the disease dies out and no epidemic takes place.

But, if $S_0 > S^*$, then, using Equation (8.5.13), $I(t)$ increases to a maximum, and then decreases to zero.

If we define the **basic reproduction number** to be

$$r_0 \equiv \frac{S_0}{S^*} = \left(\frac{\beta}{\gamma}\right)^2 S_0, \tag{8.5.15}$$

then

$$r_0 = \begin{cases} > 1, \text{an epidemic occurs,} \\ < 1, \text{no epidemic takes place.} \end{cases} \tag{8.5.16}$$

8.5.2 First-Integral

Adding Equations (8.5.1), (8.5.2) and (8.5.3) gives

$$\frac{d}{dt}(S + I + R) = 0 \Rightarrow N(t) = S(t) + I(t) + R(t) = \text{constant}, \tag{8.5.17}$$

i.e., the total population is constant.

The trajectories, $I = I(S)$, in the (S, I) plane, are solutions of the differential equation

$$\begin{aligned} \frac{dI}{dS} &= \frac{dI/dt}{dS/dt} \\ &= -1 + \sqrt{\frac{S^*}{S}}. \end{aligned} \tag{8.5.18}$$

This is a separable equation and can be readily integrafed to give

$$I(t) + S(t) - 2\sqrt{S^* S(t)} = \text{constant}. \tag{8.5.19}$$

The constant can be determined by using the following initial conditions

$$I(0) = I_0, S(0) = S_0. \tag{8.5.20}$$

Doing so, gives

$$I(t) + S(t) - 2\sqrt{S^* S(t)} = I_0 + S_0 - 2\sqrt{S^* S_0}. \tag{8.5.21}$$

This relationship is called a first-integral for Equations (8.5.1), (8.5.2), and (8.5.3).

The Figures 8.1 and 8.2 give the general behaviors of $S(t)$ and $I(t)$ for their time dependencies, and their trajectories in the (S, I) plane.

8.5.3 Maximum Infectives Number

If $S_0 < S^*$ and $I_0 > 0$, then the largest numbers of infected individuals will be I_0, i.e.,

$$S_0 < S^* \Rightarrow I_{\max} = I_0. \tag{8.5.22}$$

However, if $S_0 > S^*$, then the number of infectives will peak at $S = S^*$. From this knowledge, we can use Equation (8.5.21), with $S = S^*$ and $I = I_{\max}$, to obtain

$$S_0 > S^* \Rightarrow I_{\max} = I_0 + S_0 + S^* - 2\sqrt{S^* S_0}, \tag{8.5.23}$$

or,

$$I_{\max} = I_0 + \left(\sqrt{S_0} - \sqrt{S^*}\right)^2, \tag{8.5.24}$$

8.5.4 Final Susceptible Population Number

If $S_0 > S^*$, the situation where an epidemic occurs, then after the epidemic runs its course

$$\lim_{t\to\infty} I(t) = I_\infty = 0, \quad \lim_{t\to\infty} S(t) = S_\infty < S^* < S_0, \tag{8.5.25}$$

and we can write Equation (8.5.21) as

$$S_\infty - \left(2\sqrt{S^*}\right)\sqrt{S_\infty} - \left(I_0 + S_0 - 2\sqrt{S^*}\sqrt{S_0}\right) = 0. \tag{8.5.26}$$

This equation is quadratic in $\sqrt{S_\infty}$ and can be solved for it. After some algebraic manipulations, which use the conditions

$$S_\infty < S^* < S_0, \tag{8.5.27}$$

the following expression is obtained for S_∞,

$$S_\infty = S^*\left[1 - \sqrt{\left(\sqrt{\frac{S_0}{S^*}} - 1\right)^2 + \frac{I_0}{S^*}}\right]^2. \tag{8.5.28}$$

Note that the number of susceptibles who become ill is

$$S_{\text{ill}} = S_0 - S_\infty. \tag{8.5.29}$$

Further, the total count of the infective population is

$$I_{\text{total}} = I_0 + S_{\text{ill}} = I_0 + S_0 - S_\infty. \tag{8.5.30}$$

8.5.5 Exact, Explicit Solution

Make the following transformation of variables in Equations (8.5.1) and (8.5.2)

$$u = \sqrt{S}, \quad v = \sqrt{I}, \tag{8.5.31}$$

and obtain

$$\frac{du}{dt} = -\bar{\beta}v, \quad \frac{dv}{dt} = \bar{\beta}u - \bar{\gamma}, \tag{8.5.32}$$

where

$$\bar{\beta} = \frac{\beta}{2}, \bar{\gamma} = \frac{\gamma}{2}. \tag{8.5.33}$$

Observe that both of these first-order, differential equations are linear, with constant coefficients. Elementary substitutions give, respectively, second-order linear differential equations for $u(t)$ and $v(t)$:

$$\frac{d^2u}{dt^2} + \bar{\beta}^2 u = \bar{\gamma}\bar{\beta}, \tag{8.5.34}$$

$$\frac{d^2v}{dt^2} + \bar{\beta}^2 v = 0, \tag{8.5.35}$$

where the initial conditions are

$$u(0) = u_0 = \sqrt{S_0}, \quad \frac{du(0)}{dt} = -\bar{\beta}\sqrt{I_0}, \tag{8.5.36}$$

$$v(0) = v_0 = \sqrt{I_0}, \quad \frac{dv(0)}{dt} = \bar{\beta}\sqrt{S_0} - \bar{\gamma}. \tag{8.5.37}$$

Consequently, the solutions for $u(t)$ and $v(t)$ are

$$u(t) = \left[\sqrt{S_0} - \left(\frac{\bar{\gamma}}{\bar{\beta}}\right)\right]\cos\left(\bar{\beta}\,t\right) - \sqrt{I_0}\sin\left(\bar{\beta}\,t\right) + \left(\frac{\bar{\gamma}}{\bar{\beta}}\right), \tag{8.5.38}$$

$$v(t) = \sqrt{I_0}\cos\left(\bar{\beta}\,t\right) + \left[\sqrt{S_0} - \left(\frac{\bar{\gamma}}{\bar{\beta}}\right)\right]\sin\left(\bar{\beta}\,t\right). \tag{8.5.39}$$

Using the trigonometric relation

$$a\,\cos\theta + b\,\sin\theta = R\,\sin(\theta + \phi), \tag{8.5.40}$$

where

$$R = \sqrt{a^2 + b^2}, \tan\phi = \frac{b}{a}, \tag{8.5.41}$$

the function $v(t)$ can be rewritten to the expression

$$v(t) = \left[\left(\sqrt{S_0} - \sqrt{S^*}\right)^2 + I_0\right]^{\frac{1}{2}} \cdot \sin\left(\bar{\beta}\,t + \phi_1\right), \tag{8.5.42}$$

$$\phi_1 = \tan^{-1}\left(\frac{\sqrt{I_0}}{\sqrt{S_0} - \sqrt{S^*}}\right), \tag{8.5.43}$$

where

$$\frac{\bar{\gamma}}{\bar{\beta}} = \frac{\gamma}{\beta} = \sqrt{S^*}. \tag{8.5.44}$$

However, $I(t)$ is not the square of the above just calculated $v(t)$, i.e., $I(t) \neq v(t)^2$. We must be careful.

In more detail, consider $v(t)^2$, when $S_0 > S^*$. We know that for this case, $I(t)$ first increases, reaches a maximum value, a peak, then decreases to zero. $S(t)$ decreases monotonically from the value S_0 to its final, finite value of S_∞. Note that these general features must exist in any valid SIR model. However, $v(t)$, as given in Equation (8.5.43), oscillates and thus $v(t)^2$ also oscillates. Thus, we need to examine what $v(t)$ means in terms of $I(t)$.

At $t = 0$, $v(0) = \sqrt{I_0} > 0$ and since $dI(0)/dt > 0$, it follows that $dv(0)/dt > 0$. Combining these two results implies that after $v(t)$ reaches a maximum (positive) value, it begans to decrease and becomes zero at a time $t = t_c$. Immediately after $t = t_c$, $v(t)$ takes on negative values. Since $I(t) \geq 0$, for all $t \geq 0$, and because $v(t)$ is defined to be $v(t) = \sqrt{I(t)}$, it is clear that the function $v(t)$, expressed in Equation (8.5.43), can not represent $I(t)$ for $t > t_c$. The resolution of this issue is to note that the differential equation for $I(t)$ has also the solution $I(t) = 0$. This means that a valid piece-wise, continuous solution for $I(t)$ can be constructed such that this $I(t)$ has all of the correct features expected for the time evolution of the infective population. This solution is

$$[I(t) = \begin{cases} v(t)^2, & 0 < t \leq t_c; \\ 0, & t > t_c. \end{cases} \tag{8.5.45}$$

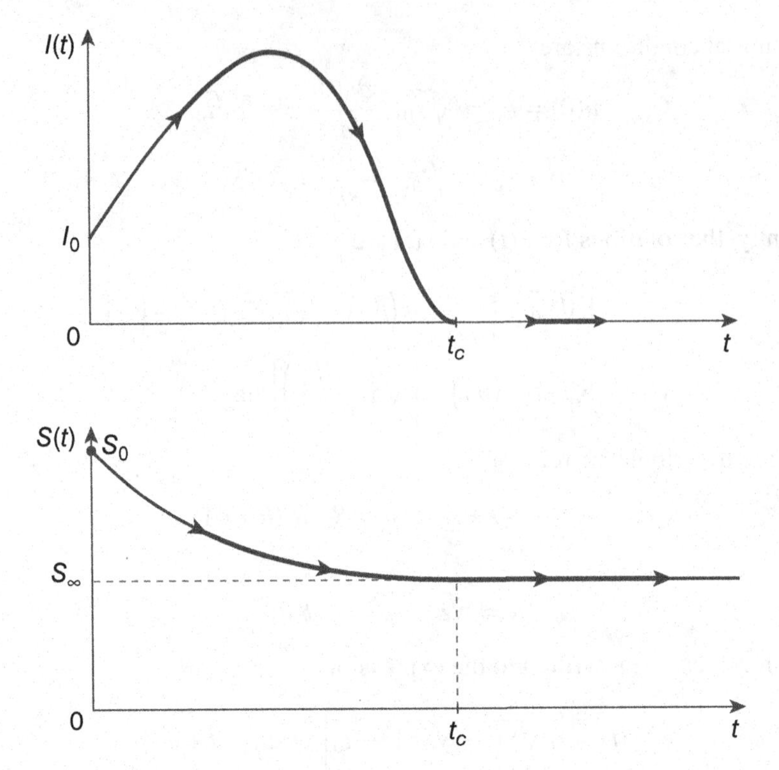

Figure 8.6: General dependencies of $I(t)$ and $S(t)$ on time.

The time t_c, where

$$I(t_c) = v(t_c)^2 = 0, \tag{8.5.46}$$

can be calculated by setting the argument of the sine function, in Equation (8.5.42), to zero and solving for t_c to obtain

$$t_c = \left(\frac{2}{\beta}\right)\left[\pi - \tan^{-1}\left(\frac{\sqrt{I_0}}{\sqrt{S_0} - \sqrt{S^*}}\right)\right]. \tag{8.5.47}$$

The solution for $S(t)$ can be determined from Equation (8.5.38) and takes the form

$$S(t) = \begin{cases} u(t)^2, & 0 < t \leq t_c; \\ S_\infty, & t > t_c, \end{cases} \tag{8.5.48}$$

where S_∞ is given in Equation (8.5.28).

A sketch of $S(t)$ and $I(t)$ is given in Figure 8.6. It shows the essential features of the time evolution of the susceptible and infective populations.

8.6 RÉSUMÉ

If one tries to modify the SIR model by the addition of other compartments, then more realist effects can be included. For example, a SIRS model includes the consequences of persons in the removed class eventually losing their immunity and going back into the susceptible population. A possible mathematical model is

$$\frac{dS}{dt} = -\beta SI + \alpha R, \tag{8.6.1}$$

$$\frac{dI}{dt} = \beta SI - \gamma I, \tag{8.6.2}$$

$$\frac{dR}{dt} = \gamma I - \alpha R, \tag{8.6.3}$$

where (α, β, γ) are positive parameters. Note that since each differential equation involves all two or more variables, this system can not be as easily investigated as the simple SIR model discussed in Sections 8.3 and 8.4. Further, it is not likely that modifications of the above equations will produce a system of differential equations for which explicit, exact solutions can be determined. However, there are three important insights that can be gleaned from the work presented in this chapter:

(i) For a given phenomena, there exist many set of equations that can provide relevant mathematical models and their associated productions. In other words, uniqueness does not exist within the realm of mathematical models for a given system.

(ii) In general, there are few **a priori** rules that can be applied to restrict the structure of mathematical models.

(iii) Mathematical modelling is hard and thought must be put into deciding just what are the fundamental concepts and issues, and how they should be translated into the mathematical equations that will then be analyzed.

PROBLEMS

Sections 8.2 and 8.3

1) The SIR model can be formulated to include population growth. The following is such a model

$$\frac{dS}{dt} = \left(aS - bS^2\right) - \beta SI,$$

$$\frac{dI}{dt} = \beta SI - \gamma I,$$

$$\frac{dR}{dt} = \gamma I.$$

Note that the total population, $N(t) = S(t) + I(t) + R(t)$, is a function of time. Since the first two equations dependent only on S and I, an analysis in the $S - I$ plane can be done. Please do so.

Section 8.4

2) Do a "flattening of the curve" analysis for the modified SIR model given in Problem 1.

Section 8.5

3) Another version of a modified SIR model with population growth is

$$\frac{dS}{dt} = a_1 \sqrt{S} - b_1 S - \beta_1 \sqrt{S} \sqrt{I},$$
$$\frac{dI}{dt} = \beta_1 \sqrt{S} \sqrt{I} - \gamma_1 \sqrt{I},$$
$$\frac{dR}{dt} = \gamma_1 \sqrt{I}.$$

(a) Find exact, explicit solutions for $S(t)$ and $I(t)$.

(b) Can $R(t)$ be calculated?

(c) If possible, calculate S_∞ and I_{max}.

NOTES AND REFERENCES

Sections 8.1, 8.2, and 8.3: Excellent introductions to SIR models and their generalizations appear in the following references

1) L. Edelstein-Keshet, **Mathematical Models in Biology** (McGraw-Hill, New York, 1988).

2) R.M. Anderson and R.M. May, **Infectious Diseases of Humans** (Oxford University, Press, Oxford, 1991).

3) H.W. Hethcote, The mathematics of Infectious diseases, **SIAM Review**, Vol. 42 (2000), 599-653.

4) F. Brauer and C. Castillo-Chavez, **Mathematical Models in Population Biology and Epidemiology** (Springer, New York, 2000).

The original work deriving the SIR methodology appear in the articles of W.O. Kermack and A.G. McKendrick; they are

5) Contributions to the mathematical theory of epidemics, **Proceedings of the Royal Society London A**, Vol. 115 (1927), 700-721; Vol. 138 (1932), 55-83; Vol. 141 (1933), 94-122.

Section 8.4: The concept of "flattening the curve" and its uses are discussed in the two references

6) Z. Feng, J.W. Glasser, and A.N. Hill, on the benefits of flattening the curve: A perspective, **Mathematical Biosciences**, Vol. 326 (August 2020), 108389.

7) C.N. Ngonghala, et al., Mathematical assessment of the impact of non-pharmaceutical interventions on curtailing the 2019 novel coronavirus, **Mathematical Biosciences**, Vol. 325 (July 2020), 108364.

As far as I am aware, the first geometric-graphic technique for illustrating and explaining the consequences of "flattening the curve" appeared in the publication

8) R.E. Mickens and T.M. Washington, The roles of SIR mathematical models in epidemiology, **Newsletter of the American Physical Society Forum on the History of Physics**, Vol. XIV, Number 5 (2020), pps. 2, 10–15.

Section 8.5: This section is a minor revision of the article

9) R.E. Mickens, An exactly solvable model for the spread of disease, **The College Mathematics Journal**, Vol. 43 (2012), 114-121.

CHAPTER 9

Dieting Model

9.1 INTRODUCTION

Dieting, generally the loss of body mass, is of concern to many individuals in societies where there is an abundance of food products laden with sugar and salt. Dieting as a process and social phenomenon is a multi-billion dollar industry dealing with the following products and services:

- diet related food products
- nutritional supplements
- sports related equipment and clothes
- exercise tapes, videos, and televisions programs
- articles, books, and other paper and electronics publications.

The main purpose of all these activities and related items is to achieve body mass reduction. However, long-term mass loss and its maintenance is very difficult for most individuals. A broad range of complex personal and physiological facts influence this situation.

The time line for most individuals, who attempt to lose weight, generally, goes as follows:

(i) A decision is made to reduce body-mass by a certain amount.

(ii) A diet and physical exercise plan is choosen.

(iii) Initially, there is rapid mass loss, followed by modest decreases.

(iv) Eventually, the desired mass is attained.

(v) This mass lost is maintained for a generally short interval of time.

(vi) During this interval, the individual returns to essentially the same "eating pattern" as they had prior to the start of the diet.

(vii) Rather quickly, their body-mass increases and reaches a body-mass that may be larger than the pre-diet mass.

(viii) For many, the reaction is to repeat (i), again.

DOI: 10.1201/9781003178972-9

The main purpose of this chapter is to construct a rather straightforward mathematical model of the dieting process and investigate the properties of its solutions. This model consists of a single, nonlinear, first-order differential equation. Unfortunately, as is the case for almost all differential equation, no exact, explicit solution exists in terms of a finite combination of the elementary functions. However, this does not prevent us from determing all of the important qualitative aspects of the solutions, along with several of its quantitative features. In particular, use is made of phase space (geometrical) methods and linear stability analysis. Further, we construct two ansatzes for approximations to the actual analytical solutions.

The significance of this model is that it provides important insights and guides to a biomedical issue that is of interest to the general public. We make no attempt to estimate parameters or to fit data. This feature illustrates that the value of a mathematical model can come just from the knowledge gained from its derived qualitative properties.

Section 9.2 presents our dieting mathematical model and the assumptions on which it is based. We also determine, using dimensional analysis, the characteristic time scale of the system. In Section 9.3, we analyze, in detail, the general properties of the solutions to the model. We show in Section 9.4 how to construct two different ansatzes for approximations to the exact solutions and use one to estimate the time to lose a fixed amount of body-mass. Finally, in Section 9.5, we summarize the results from using this model of dieting.

9.2 MATHEMATICAL MODEL

9.2.1 Assumptions

To start, we make the following definitions:

M(t): body-mass of the individual at time t;

F(t): food/nutrition intake functions at time t;

E(t): function that characterizies the physical activity of the individual at time t;

W(M): body metabolism function which depends only on $M(t)$.

Note that $F(t)$ is a non-negative function which gives a measure of the caloric intake of the dieter. $E(t)$ is also a non-negative function and correlates with the non-metabolic energy activities of the body, with physical exercise being the dominant component.

The function, $W(M)$, is non-negative and all the data are consistent with the form

$$W(M) \sim M^{\alpha}, \quad \alpha \approx 0.7. \tag{9.2.1}$$

For our purposes, we take for $W(M)$ the function

$$W(M) = \beta M^{3/4}, \tag{9.2.2}$$

where β is a positive constant. A more general and complex form for $W(M)$ is

$$W(M) = \frac{AM}{1 + BM^{(1-\alpha)}}, \quad 0 < \alpha < 1, \tag{9.2.3}$$

where A and B are positive constants.

Our dieting model is

$$\frac{dM(t)}{dt} = F(t) - E(t) - W(M), \tag{9.2.4}$$

with $W(M)$ selected to be the function given in Equation (9.2.2), i.e.,

$$\frac{dM(t)}{dt} = F(t) - E(t) - \beta M(t)^{\frac{3}{4}}. \tag{9.2.5}$$

This is a first-order, nonlinear, time-dependent differential equation which has no known, exact, explicit solution.

A critical issue is what should be the functions $F(t)$ amd $E(t)$?

9.2.2 Simplification Condition

A diet consists of selecting a strategy for the determination of the amount of food (calories) consumed during a fixed interval of time, along with a similar determination of the calories to be expended in directed physical activity during the same interval of time. The simplest and most direct strategy is to take both the food/nutrition intake and the exercise regiment to be "constant" over a convenient and meaningful time interval. For example, the caloric intake can be chosen to be $\bar{F}$ calories/day from food/nutrition, with $\bar{E}$ calories/ day in exercise, where $\bar{F}$ and $\bar{E}$ are assumed to be constant. Note that the reference time interval is one day.

This diet strategy leads to the result

$$\begin{aligned} [F(t) - E(t)]_{\text{average}} &= \bar{F} - \bar{E} \\ &\equiv \lambda \end{aligned} \tag{9.2.6}$$

where the item in the square-brackets is averaged over one day. In general, under normal circumstances, λ will be positive. With this dieting strategy, Equation (9.2.5) takes the form

$$\frac{dM}{dt} = \lambda - \beta M^{\frac{3}{4}}. \tag{9.2.7}$$

In the work and analysis to follow, we do not need to know the actual values of two parameters λ and β. Our interest is only in the general qualitative features of the solutions to Equation (9.2.7).

9.2.3 Time Scale

One aspect of any model of physical phenomena is its time scale, T^*. This value, for our case, gives the number of days it takes the individual to make a significant change in their body-mass.

This parameter may be estimated by using dimensional analysis. To do so, express T^* as

$$T^* = C\lambda^a \beta^b, \tag{9.2.8}$$

where C is a dimensionless constant of order one. If the physical unit of M is denoted by #, i.e.,

$$[M] = \#, \tag{9.2.9}$$

then

$$[\lambda] = \frac{\#}{T}, \quad [\beta] = \frac{(\#)^{\frac{1}{4}}}{T}, \tag{9.2.10}$$

where T = time unit. In Equation (9.2.8), a and b are constants to be determined. Substituting these results in Equation (9.2.8) gives

$$[T^*] = [C][\lambda]^a[\beta]^b \tag{9.2.11}$$

and

$$T = [1]\left(\frac{\#}{T}\right)^a\left[\frac{(\#)^{\frac{1}{4}}}{T}\right]^b, \tag{9.2.12}$$

and finally,

$$T = (\#)^{a+\frac{b}{4}}(T)^{-a-b}. \tag{9.2.13}$$

Comparing both sides of Equation (9.2.13) produces the result

$$a + \frac{b}{4} = 0, \quad -a - b = 0. \tag{9.2.14}$$

Solving for a and b gives

$$a = \frac{1}{3}, \quad b = -\frac{4}{3}. \tag{9.2.15}$$

If we take $C = 1$, then $T*$ is

$$T^* = \frac{\lambda^{\frac{1}{3}}}{\beta^{\frac{4}{3}}} = \left(\frac{\lambda}{\beta^4}\right)^{\frac{1}{3}}. \tag{9.2.16}$$

9.3 ANALYSIS OF MODEL

The body-mass, dieting Equation (9.2.7) has the constant solution

$$M^* = \left(\frac{\lambda}{\beta}\right)^{\frac{4}{3}}. \tag{9.3.1}$$

Since

$$\frac{dM}{dt} = \begin{cases} < 0, & M > M^*, \\ > 0, & M < M^*, \end{cases} \tag{9.3.2}$$

it follows that for

$$M(0) = M_0 > 0, \tag{9.3.3}$$

the solution decreases to the value M^* if $M_0 > M^*$, while it increases to M^* if $M_0 < M^*$. Note that $M_0 = M^*$, then $M(t)$ remains at this value. In any case, we have

$$\lim_{t\to\infty} M(t) = M^*, M_0 > 0. \tag{9.3.4}$$

This result implies that the constant solution $M(t) = M^*$ is globally stable, i.e., regardless of where one starts, for $M(0) = M_0 > 0$, the value of the body-mass ends up at the mass M^*. See Figure 9.1

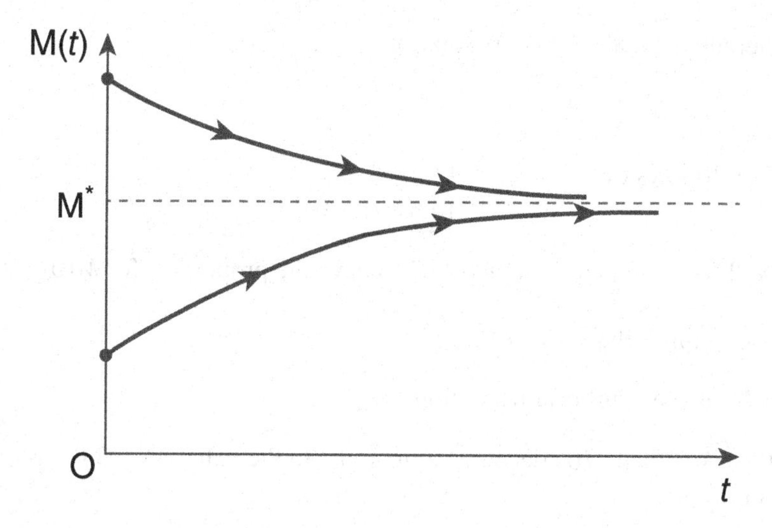

Figure 9.1: Typical solutions of the body-mass differential equation for $M_{(0)} > 0$.

While Equation (9.2.7) can not be solved exactly, its solutions close to the constant solution (fixed point or equilibrium value) can be determined. To do this, take the following form for $M(t)$

$$M(t) = M^* + \epsilon(t), \tag{9.3.5}$$

where

$$|\epsilon(0)| \ll M^*. \tag{9.3.6}$$

The perturbation $\epsilon(t)$ can take either sign. Substituting this expression into Equation (9.2.7) gives

$$\frac{d\epsilon}{dt} = \lambda - \beta[M^* + \epsilon]^{\frac{3}{4}}. \tag{9.3.7}$$

Expanding the item in the square bracket produces the result

$$\begin{aligned} [M^* + \epsilon]^{\frac{3}{4}} &= \left[M^*\left(1 + \frac{\epsilon}{M^*}\right)\right]^{\frac{3}{4}} \\ &= (M^*)^{\frac{3}{4}}\left(1 + \frac{\epsilon}{M^*}\right)^{\frac{3}{4}} \\ &= (M^*)^{\frac{3}{4}}\left[1 + \left(\frac{3}{4}\right)\left(\frac{\epsilon}{M^*}\right) + \cdots\right]. \end{aligned} \tag{9.3.8}$$

If only the linear term is retained, then after some algebraic manipulations, Eg. (9.3.7) becomes

$$\frac{d\epsilon}{dt} = -R\epsilon, \quad \epsilon(0) = \epsilon_0, \tag{9.3.9}$$

where

$$R = \left(\frac{3}{4}\right)\left[\frac{\beta}{(M^*)^{\frac{1}{4}}}\right] = \left(\frac{3\beta}{4}\right)\left(\frac{\beta}{\lambda}\right)^{\frac{1}{3}}, \tag{9.3.10}$$

and this differential equation has the solution

$$\epsilon(t) = \epsilon_0 e^{-Rt}. \tag{9.3.11}$$

Since $R > 0$, it follows that

$$\lim_{t\to\infty} \epsilon(t) = 0. \tag{9.3.12}$$

In summary, the solutions to Equation (9.2.7) have the properties, for $M(0) > 0$:

(1) $M(0) > 0$, implies that $M(t) > 0$.

(2) $M(t) = M^*$ is the equilibrium solution.

(3) If, $M(0) > M^*$, then $M(t)$ deceases monotonic to the value M^*. This case corresponds to lost of mass.

(4) If, $M(0) < M^*$, then $M(t)$ increases monotonic to the value M^*. We now have gain of mass.

The conclusions reached, in (1) to (4), are made under the assumption that the parameters λ and β are held constant. In practice, β is generally determined by the individuals body physiology and will not change. However, λ is under the direct control of the individual.

9.4 APPROXIMATE SOLUTIONS

While our mathematical dieting model equation

$$\frac{dM}{dt} = \lambda - \beta M^{\frac{3}{4}}, \; M(0) > 0, \tag{9.4.1}$$

can not be solved exactly, in terms of a finite combination of elementary functions, approximations to its solutions can be constructed. At minimum any approximation should satisfy the four conditions or features listed near the end of the previous Section 9.3. We now write down an analytic approximation to the solutions of Equation (9.4.1). It takes the form

$$M_a(t) = A - Be^{-Rt}, \tag{9.4.2}$$

where R is given by Equation (9.3.10) and the constants A and B are to be determined from knowledge of

$$M(0) = M_0, \; M(\infty) = M^*. \tag{9.4.3}$$

Imposing the conditions in Equation (9.9.3) on $Ma(t)$ gives

$$M_0 = A - B, \; M(\infty) = A. \tag{9.4.4}$$

Therefore, $M_a(t)$ is

$$M_a(t) = M^* - (M^* - M_0)\, e^{-Rt}. \tag{9.4.5}$$

An inspection of Equation (9.4.5) shows that this ansatz reproduces all four of the conditions stated near the end of Section 9.3.

Another ansatz is the function

$$M_a(t) = \frac{A_1}{1 + B_1 e^{-Rt}}, \tag{9.4.6}$$

where R is the value stated in Equation (9.3.10). Imposing the conditions of Equation (9.4.3) gives

$$M_0 = \frac{A_1}{1 + B_1}, \; M^* = A_1 \,, \tag{9.4.7}$$

and

$$A_1 = M^*, B_1 = \frac{M^* - M_0}{M_0} \,. \tag{9.4.8}$$

Therefore, the second ansatz is

$$M_a(t) = \frac{M^* M_0}{M_0 + (M^* - M_0)\, e^{-Rt}}. \tag{9.4.9}$$

Suppose the dieter wants to lose mass. Let

- λ and β be known.
- Assume, for these values that currently

$$M_0 > M.$$

The following question can be asked:

How long does it take to achieve a mass lost of ΔM, where

$$\Delta M = M_0 - M_1, \qquad M_0 > M_1 > M^*?$$

This time interval can be estimated from the knowledge of the approximate solution, $M_a(t)$. To proceed, let us use $M_a(t)$ as expressed in Equation (9.4.5).

Let t_1 be the time interval from M_0, at $t = 0$, to $M_1 = M(t1) \simeq M_a(t_1)$. Therefore, we have

$$M_a\,(t_1) = M^* - (M^* - M_0)\, e^{-Rt_1}, \tag{9.4.10}$$

$$M_{a|}(t_1) = M_1. \tag{9.4.11}$$

Therefore,

$$M_1 - M^{(*)} = (M_0 - M^*)\, e^{-Rt_1}, \tag{9.4.12}$$

and

$$e^{Rt_1} = \left(\frac{M_0 - M^*}{M_1 - M^*}\right) > 1. \tag{9.4.13}$$

Solving for t_1, gives

$$t_1 = \left(\frac{1}{R}\right) \text{Ln}\left(\frac{M_0 - M^*}{M_1 - M^*}\right). \tag{9.4.14}$$

Note that R has the physical units of inverse time. From the definition of R, we have the following connection between R and the characteristic time, T^*, see Equation (9.2.16),

$$\frac{1}{R} = \left(\frac{4}{3}\right) T^*. \tag{9.4.15}$$

9.5 DISCUSSION

There exists in the research literature a large number of mathematical models for dieting. We have presented an elementary model which incorporates all of the major features of the dieting process. This model

$$\frac{dM}{dt} = \lambda - \beta M^{\frac{3}{4}}, \tag{9.5.1}$$

depends only on two parameters; λ, which measures the impact of food intake and exercise, i.e.,

$$\begin{aligned} \lambda &\equiv [\text{Food} - \text{Exercise}]_{\text{average}} \\ &= \bar{F} - \bar{E}, \end{aligned} \tag{9.5.2}$$

and β, which is related to a particular individual's basic internal metabolism. The "dieting strategy" is characterized by selecting a (positive) value for λ. Once λ is known, then the equilibrium mass is determined by the relationship

$$M^* = \left(\frac{\lambda}{\beta}\right)^{4/3}. \tag{9.5.3}$$

It should be noted that for a given value of λ, there are many combinations of $\bar{F}$ and $\bar{E}$ which can satisfy this condition. In other words, a given dieting strategy can be achieved by various choices of caloric (food) restriction and physical exercise.

Further examination of this model shows that the largest decrease of body-mass per unit time occurs at the beginning of the diet. Similarly, the smallest body-mass loss takes place when the weight goal is nearly accomplished. An important result, not generally stated in the analysis of dieting models, is that to remain at the desired body-mass, M^*, the individual must maintain the diet strategy that produced the mass loss "forever." All the consequences of our model are consistent with the current data on dieting.

Finally, we end this chapter with a restatement of the dieting process, which includes both body-mass loss and gain.

Body-Mass Loss

(i) Denote the initial body-mass by M_i^* and assume that a smaller body-mass is the goal of the dieter, M_f^*, i.e.,

$$M_f^* < M_i^*. \tag{9.5.4}$$

(ii) To obtain this reduction, the parameter λ must change from λ_i to λ_f, where

$$\lambda_i = \beta(M_i^*)^{\frac{3}{4}} > \lambda_f = \beta\left(M_f^*\right)^{\frac{3}{4}}. \tag{9.5.5}$$

(iii) The initiation of this new dieting strategy will result in the body-mass lost of

$$\Delta M = M_i^* - M_f^*. \tag{9.5.6}$$

See Figure 9.2,

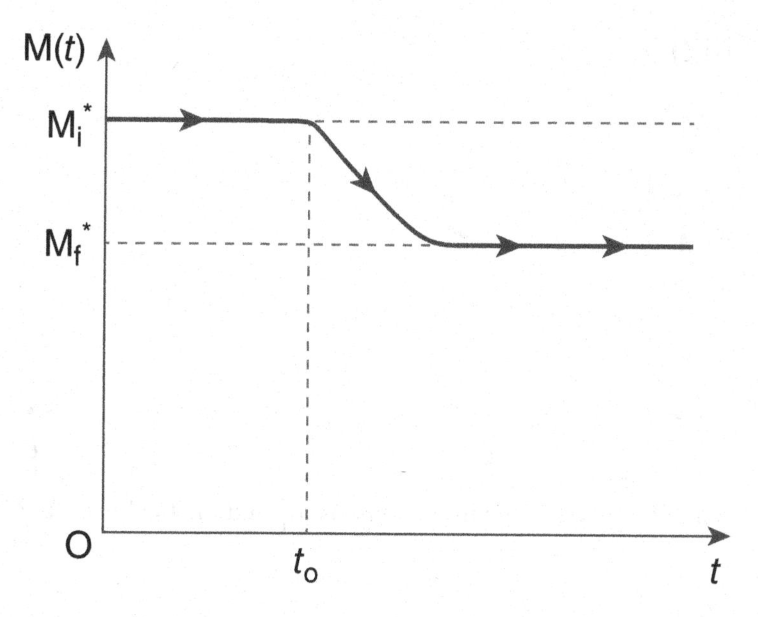

Figure 9.2: Loss of body-mass. The initial mass is M_i^* and the final mass is M_f^*, with $M_f^* < M_i^*$.

Body-Mass Gain

This situation is the reverse of the previous case. Here the individual must increase the value of λ, i.e.,

$$\lambda_i < \lambda_f. \tag{9.5.7}$$

If the two body-masses are M_i^* and M_f^*, where M_f^* is the final, larger mass, then λ_f should be selected to have the value

$$\lambda_f = \beta\left(M_f^*\right)^{\frac{3}{4}}. \tag{9.5.8}$$

The resulting increase in body-mass is depicted in Figure 9.3.

PROBLEMS

Sections 9.2 and 9.3

1) Consider the following linear mathematical model for the dieting process

$$\frac{dM}{dt} = \lambda_1 - \beta_1 M, \quad M(0) = M_0 > 0.$$

Note that it can be solved exactly. Determine it major qualitative features and compare them to the model proposed in this section, i.e.,

$$\frac{dM}{dt} = \lambda - \beta M^{\frac{3}{4}}.$$

Carry out a similar analysis for the model

$$\frac{dM}{dt} = \lambda_2 - \frac{\beta_2 M}{1 + aM^{\frac{1}{4}}}, \quad a > 0.$$

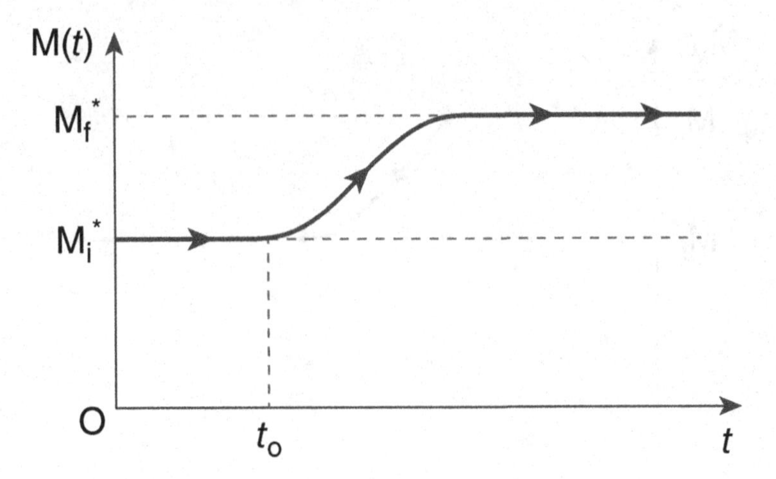

Figure 9.3: Gain of bodymass. The initial mass is M_i^* and the final mass is M_f^* with $M_i^* < M_f^*$.

What are the time scales and equilibrium values for this system?

Section 9.4

2) Justify using the approximation to the solution

$$M_a(t) = A - Be^{-Rt}, \tag{9.5.9}$$

for all three mathematical models given in Problem (1). Will R be the same for each model?

Section 9.5

3) Since the approximate solution ansatzes have exponential decaying terms, the time interval to go from body-mass M_i^* to M_f^* is infinite. However, this is a mathematical issue and not an actual physical problem. In reality, this time has a finite value. Think about this situation and resolve it within the context of how actual physical measurements are made.

NOTES AND REFERENCES

Section 9.1: There are a plethora of a dieting related mathematical models and an internet search will bring them to easily view. The following three articles are examples of what is available

1) D.M. Thomas, M.C. Gonzalez, A.Z. Pereira, L.M. Redman, and S.B. Heymsfield, Time to correctly predict the amount of weight loss with dieting, **Journal of the Academy of Nutrition and Dietics**, Vol. 114 (2014), 857-861.

2) K.D. Hall, D.A. Schoeller, and A.W. Brown, Reducing calories to lose weight, **Journal of the American Medical Association**, Vol. 319 (2018), 2336-2337.

3) D.M. Thomas, M. Scioletti, and S.B. Heymsfield, Predictive mathematical models of weight loss, **Current Diabetes Reports**, Vol. 19 (2019), Article number 19.

Section 9.2 and 9.3: This section is based on the article

4) R.E. Mickens, D.N. Brewley, and M.L. Russell, A model of dieting, **SIAM Review**, Vol. 40 (1998), 667-672.

The theoretical reasoning behind the use of the form $M^{3/4}$ is presented in the article

5) G.B. West, J.H. Brown, and B.J. Enquist, A general model for the origin of allometric scaling laws in biology, **Science**, Vol. 276 (1997), Issue 5309, 122-126.

CHAPTER 10

Alternate Futures

10.1 INTRODUCTION

This chapter attempts to model certain aspects of human interactions and experiences with the physical universe. In particular, we introduce and examine the concepts of "alternative futures (AFs)" and "counterfactual histories (CFHs). The essential purpose of the chapter is to give arguments for the proposition that alternative futures can exist. For a system to have AFs, it must have the feature that at some time, $t = t_0$, its future evolution should manifest itself such that more than one outcome is (physically) possible. These multi-outcomes are a consequence of the system interacting with its environment. However, before providing a clear and fuller exploration of these topics and related issues, we illustrate their occurrence in two elementary, but not simple, model systems, namely,

- coin flipping,
- going from location A to location B.

The following two sections discuss these systems.

Section 10.4, summarizes the major general results coming from the two model systems investigated in Sections 10.2 and 10.3. A number of critical definitions are given and briefly discussed. The final, Section 10.5, gives arguments that connect AFs to the idea that CFHs must exist, i.e., any system possessing AFs must also give rise to the existence of CFHs.

A very interacting, but important issue to note is that the presented general arguments do not depend on whether the systems satisfy classical or quantum physical laws.

10.2 TWO SYSTEMS EXHIBITING ALTERNATIVE FUTURES

10.2.1 Coin Flipping

One of the simplest physical systems is one that has only two states. Such a system can be modelled by a two-sided coin. In more detail, we take the system to be composed of the following components:

- a two-sided coin, with one side labeled H, the other labeled T;

DOI: 10.1201/9781003178972-10

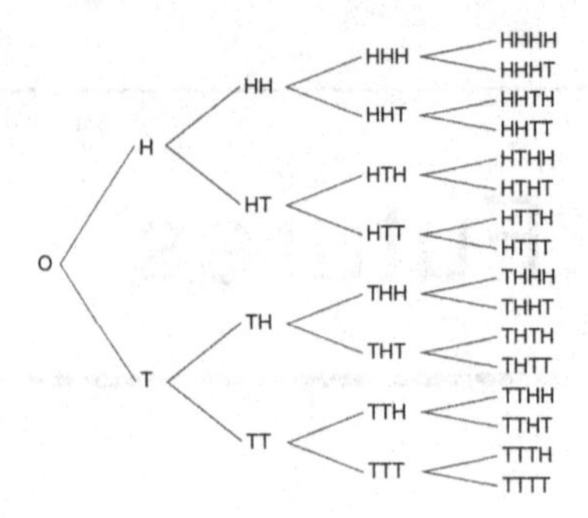

Figure 10.1: Outcomes of flipping a coin four times.

- a mechanism to flip the coin;
- a devise to record the listing of H's and T's after a given sequence of coin flippings.

An **outcome tree** is a diagram or figure giving the possible sequences of H's and T's. Figure 10.1 is the outcome tree for four coin flippings. It is very important to observe that the ordering is critical, i.e.,

HT: means that the first flip was H and the second flip was T,

TH: means that the first flip was T, while the second flip was H.

A **branch** or **trajectory** on the outcome tree gives a particular sequence of flippings. For example, in Figure 10.1, we represent all of the possible branches after four (4) flips of the coin, starting from the unflipped state O. The sixteen distinct possibilities are indicated by the labelings on the right-side of the diagram. In general, after N flips, 2^N trajectories exist. This result is a consequence of the fact that each flip creates two possible outcomes, either an H or a T.

It is important to understand that each of these sequences is physically realizable, i.e., a flipping of an actual coin can reproduce any one of the indicated sequences. A detailed inspection of Figure 10.1, which corresponds to four flips of the coin, allows the following conclusions to be drawn:

(1) Every entry listed in the last column has a unique trajectory linking it to the initial state, O. For example, consider HTTH. It came from the prior state HTT, which came from HT, which in turn had its genesis in the state H, and his state was a consequence of a coin flip at the initial state O.

(2) The exact state, after one flipping, anywhere in the sequence, does not allow at that point a prediction of the state at any future flippings. To illustrate this, consider the state TH. After two additional flippings, the following four distinct possibilities exist: THHH, THHT, THTH, and THTT.

Reflection on this matter indicates that these results can be generalized to allow the following conclusions to be reached:

(i) From the initial, unflipped state O, after N flips. 2^N sequences are possible.

(ii) All of the 2^N sequences are realizable in the physical sense that they could be the outcome of an actual sequence of coin flippings.

(iii) A given state or sequence after k-flips has a unique trajectory going backwards to the unflipped state O. We can call this the history of this specific state at the k-th level.

(iv) A particular state, i.e., a definite sequence of H/T values at the k-th level of flipping, does not allow the exact prediction of its state (sequence) after an additional M-flips. This is a consequence of the fact that the additional M-flippings give rise to 2^M further possible states. We can label these new states, **alternative futures** of a state created by k-flips.

(v) If two separate trajectories intersect, then up to the point where they intersect, they both have exactly the same history, but they will not ever intersect again in their alternative futures.

In summary, from a given state, its past or history is exactly known, while its future has alternative possibilities.

10.2.2 Going from A to B

A second example of an elementary system having alternative futures is one associated with moving or traveling from a location A to another location B. This particular system has the following components.

(a) The different motions or movements occur within the confines of a bounded space.

In Figure 10.2, the outer rectangle is the confining boundary, while the smaller shaded rectangles are barriers to movement, with motion restricted to the pathways such as

$$5 \leftrightarrow B, 1 \leftrightarrow A, \leftrightarrow 2, 4 \leftrightarrow 3$$
$$4 \leftrightarrow 1 \leftrightarrow 5, 3 \leftrightarrow 2 \leftrightarrow B.$$

The double arrows imply that movement can take place in either direction along the pathway.

(b) The task is to leave position A, at same time t_0, and arrive at position B at a later time.

Note that the time of arrival at B is not needed. Also, restrictions such as the shortest time are not placed on the movement.

Consider the possible paths between A and B. The following is a subset of these paths:

$$A \rightarrow 2 \rightarrow B$$
$$A \rightarrow 1 \rightarrow 5 \rightarrow B$$
$$A \rightarrow 1 \rightarrow 4 \rightarrow 3 \rightarrow 2 \rightarrow B$$
$$A \rightarrow 2 \rightarrow 3 \rightarrow 4 \rightarrow 1 \rightarrow 5 \rightarrow B.$$

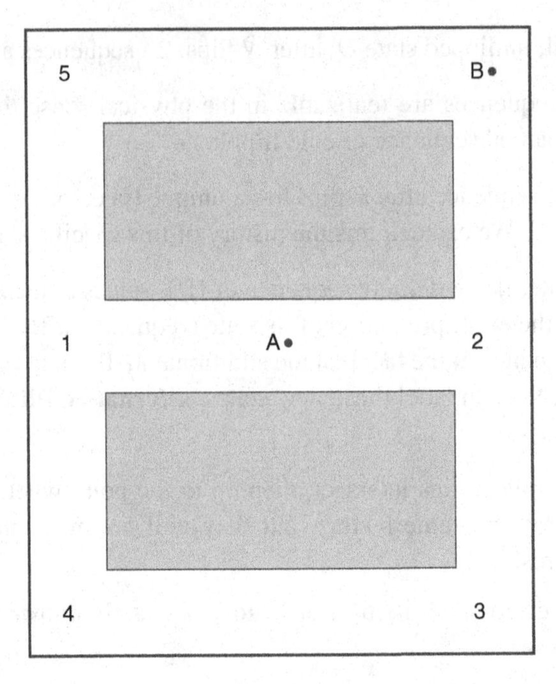

Figure 10.2: The goal is to travel from A to B. The numbers indicate possible locations that can be a part of a specific path.

If doubling-back is permitted, then very long paths exist; for example,

$$\begin{aligned} A &\to 1 \to A \to 2 \to 3 \to 4 \to 1 \to 5 \to 1 \\ &\to 5 \to 1 \to 5 \to B. \end{aligned}$$

It is clear that multi-paths or trajectories exist to get from A to B. Further, once an individual selects a given path, they will have no knowledge or experience of the paths not taken. Thus, while this system is somewhat more complex than the flipping of a coin system, the basic ideas and related concepts still hold, i.e., for a given path, there is a unique trajectory back to A from B. This means that for the path

$$A \to 1 \to 4 \to 3 \to 2 \to B,$$

an individual will experience or remember this path, but not an alternative one, such as

$$A \to 1 \to 5 \to B.$$

Keep in mind that all of the possible paths are physical realizable.

10.3 RÉSUMÉ OF CONCEPTS AND DEFINITIONS

For additional clarity, we introduce and discuss briefly, several old and new concepts and ideas presented in Section 10.2.

System: In general, the nature of a particular system and its components are readily defined and understood. For example, the two systems investigated in Section 10.2 are easily defined and clearly understood in terms of how they are defined.

States of a System: These items readily follow from how the system is defined. For the flipping of a coin situation, there are two states for a given coin, namely, head (H) and tail (T). Likewise, for one die, there are six states: the faces have from one to six dots or spots displayed on them, with no repeats.

Interaction: This is a process that allows a system to change its current state or (sometimes) to remain in that state. For the coin flipping system, if the initial state is H, then the flip (i.e., interaction) will either change H $\rightarrow$ T or produce no change, i.e., H $\rightarrow$ H.

History of a System: This is the sequence of prior states that a system underwent to arrive at its current state.

Trajectory of a System: The "trajectory of a system" and the "history of a system" are closed connected. The trajectory of a system may also include the possible alternative future sequencies.

Realizable System: A system is realizable if it obeys all of the laws of physics and can actually exist in the physics universe.

Alternative Futures: If a system undergoes an interaction, at some time $t = t_1$, such that multi-outcomes are possible, then alternative futures are said to exist. In other words, at $t = t_1$, the original (unique) trajectory **bifurcates into multi-trajectories**. Consequently, for $t > t_1$, the system has multi-sequences of possible future states.

10.4 COUNTERFACTUAL HISTORIES

A major goal of this chapter is to examine several simple realizable physical systems and demonstrate that they exhibit the phenomena of alternative futures. This was done in Section 10.2.

We now turn to the subject of counterfactual histories. Counterfactual histories are defined as possible histories of a system that were not in fact actualized. Within the framework of this chapter, counterfactual histories may be defined or conceptualized in the following manner:

(i) Consider a physical realizable system that can interact with its "environment" to produce multi-outcomes at each interaction.

(ii) If the interaction takes place at time $t = t_1$, then an "observer" of this system will have experienced a unique history for the system, and further, will experience only one of the future possibilities for the system for $t > t_1$.

(iii) The various possible multi-outcomes will be called or labeled alternative futures of the
system.

(iv) This logic or line of reasoning leads to the conclusion that any system exhibiting alternative futures will also have the feature of possessing alternative histories.

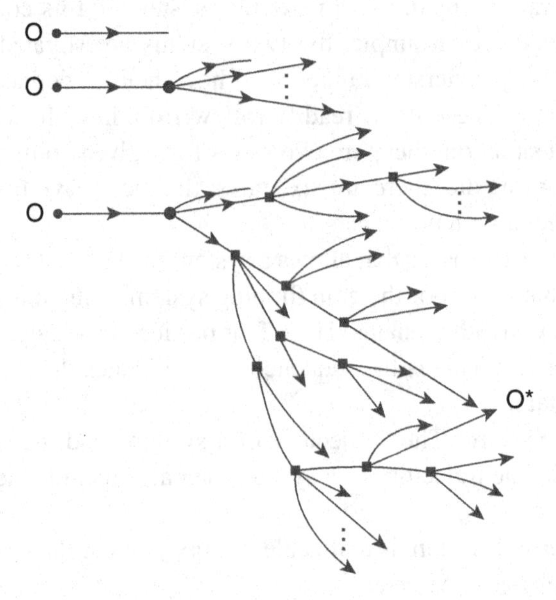

Figure 10.3: The general evolution of a system O. The dots indicate events where a system interacts with its environment and creates the possibility for multi-outcomes.

To demonstrate this phenomena, consider Figure 10.3. Starting with a system at O, after each interaction, the system can have multi-outcomes, the number of which need not be a priori specified. The only requirement is that it be at least two. In the general, the value of the "splittings" may change from one interaction to another. The details will not influence or change the final conclusions.

Consider Figure 10.3; it shows a system starting at O and then undergoing splitting at various future times. (The horizontal direction, left to right, indicates increasing time.) The top illustration shows the evolution of the system with no change taking place. The middle illustration shows the system after one splitting, while the bottom illustration is a résumé of some of the possible future splittings. Observe that from O many possible future branches exist. However, given a location on a future branch, there is only reversed path back to O. To get a "feel" for the last statement, consider a future state O^*, and verify that there is only one backward traveling trajectory to O. The same result holds for any other future state of the system. In summary:

(a) Every initial system has a multitude of possible future states, i.e., they have the property of having alternative futures.

(b) From any given future state, O^*, there is a unique backward path to O, i.e., each state O^* has a unique history.

(c) An additional consequence is that the number of future states increases at least exponentially. For the case of flipping a coin, after N flippings, the number of possible states is 2^N.

Translating these results into the realm of human experiences, (a) and (b) are consistent with the facts that each person has a unique history, but an unknown future.

(d) The existence of multi-alternative futures for a system implies the existence of counterfactual histories.

The restrictions of valid physical theories, whether classical or quantum, cannot be used to show that counterfactual histories do not exist. But the existence of a phenomenon, within the framework of a physical theory, is not the same as being able to experience all manifestations of that phenomenon. A simple illustration is provided by the flipping of a coin. After seven flips each of the following sequences are possible

HHTHTTT, HTTHTTH, THTHHTH, etc.

however, only one will actually occur and be within the experience of the flipper. At the end of the seven flips, the flipper will know exactly the outcome of the flips, but will be uncertain as to the outcomes of future flips.

The argument of the last paragraph implies that the concept of alternative futures, which implies the existence of alternative futures, which then implies the existence of counterfactual histories, is not scientific in the sense of being directly verifiable (Simonyi 2012: Tegmark 2014). However, not valid all knowledge and the understandings which come with it, is or has to be scientific (Boghossian 2007). Nonscientific (nonverifiable) knowledge can give insights into the nature of knowledge in general, while also providing constraints on what is or is not scientific knowledge. The only requirement is that the analyses must be logically consistent.

10.5 SUMMARY AND DISCUSSION

After examing two elementary systems, which are realizable both in practice and theory, we have given arguments in favor of the actual existence of the concepts "alternative futures" and "counterfactual histories." Within the presented framework, it follows that

$$\text{Counterfactual Histories} \Longleftrightarrow \text{Alternative Futuries.}$$

As stated, our systems and their evolution can be easily characterized within the framework of classical physical theory (Simonyi, 2012). However, there also exists a large body of research and discussion on related issues arising in the interpretation of certain features in quantum physics (Carr, 2007; Osnaghi et al., 2009; Tegmark, 2014). In addition, further scholarship and analysis has been done within the disciplines of history (Bunzi, 2004; Carr, 2001), historical dynamics (Turchin, 2003), and the genre of alternative history (McKnight, 1994).

We end this chapter with (another) illustration of alternative futures and counterfactual histories.

In Figure 10.4, the horizontal direction indicates time, with increasing time going from left to right. The trajectory of the system starts at O and spits seven times to get to the present, labeled "7." Note that there is only one path or trajectory from O to 7, and only it has been drawn. At point 7, the system continues to evolve, with additional splittings. Careful inspection of this diagram allows the following conclusions to be reached:

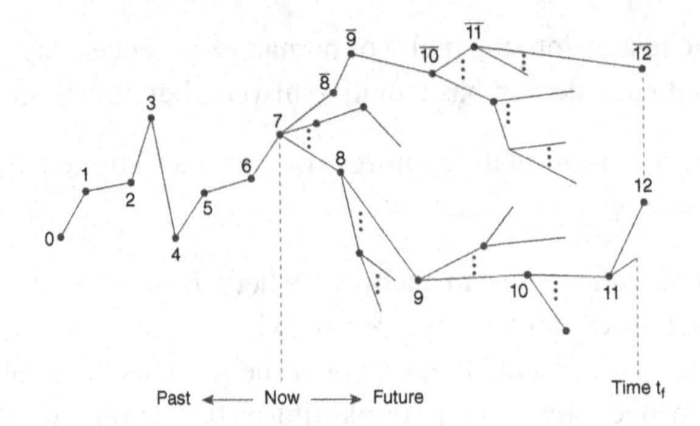

Figure 10.4: Illustration of alternative futures and counterfactual histories.

(I) The state or location 7 has a unique trajectory tracking back to O, i.e., it has a unique history of evolution from the state O.

(II) From state 7, the system has many alternative future trajectories. In particular, there are the two paths

$$7 \to 8 \to 9 \to 10 \to 11 \to 12$$
$$7 \to \overline{8} \to \overline{9} \to \overline{10} \to \overline{11} \to \overline{12}.$$

If both states 12 and $\overline{12}$ occur at the same time, then it easy to understand that these two different trajectories are alternative futures relative to the state 7.

(III) Relative to each other, the trajectories leading to states 12 and $\overline{12}$, are counterfactual histories of each other.

It is clear that more work needs to be done on the issues raised in this chapter. One such issue is to determine the overall logical consistency of these ideas. Needless to say, this is a contentious area of scholarship and research.

PROBLEMS

Section 10.2

1) Consider a system composed of a single, two sided coin and a six sided die. What is the total number of distinct states? After k simultaneous flips of the coin and die, how many possible states can the system be in? Does this system exhibit the phenomena of alternative futures and counterfactual histories?

Sections 10.4 and 10.5

2) Discuss the value or not of using the concepts of alternative futures and counterfactual histories to the discipline of history.

3) How have the general themes presented in this chapter been applied to provide resolutions to several issues arising in quantum physics? What conclusions were reached?

NOTES AND REFERENCES

Section 10.2 and 10.3: The discussion of this chapter is a revised and extended version of the following publications

1) R.E. Mickens and C. Patterson, Systems exhibiting alternative futures, **Georgia Journal of Science**, Vol. 75, No. 2 (2017), Article 4.

2) C. Patterson and R.E. Mickens, Counterfactual history is consistent with physics, **Bulletin of the Amercian Physical Society**, Vol. 61, No. 19 (2016), Abstract D4.00005.

For a quick, general overview of the concept of counterfactual history, see the website

3) **Wikipedia:** "Counterfactual history."

   Another very useful article is

4) R.N. Lebow, Counterfactual thought experiments: A necessary teaching tool, **The History Teacher**, Vol. 40, No. 2 (February 2007), 153–176.

Sections 10.4 and 10.5: Several useful and quite interesting articles and books discussing a range of topics related to this chapter are

5) M. Bunzi, Counterfactual history: A user's guide, **American Historical Review**, Vol. 109 (2004), 845–858.

6) B. Carr, editor, **Universe or Multiverse**? (Cambridge University Press, Cambridge, 2007).

7) E.V. McKnight, **Alternative History: The Development of a Literary Genre** (University of North Carolina Press; Chapel Hill, NC).

8) S. Osnaghi, F. Freitas, and O. Freire, Jr., The origin of the Everettian heresy, **Studies in the History and Philosophy of Modern Physics**, Vol. 40 (2009), 97–123.

9) K. Simonyi, **A Cultural History of Physics** (CRC Press; Boco Raton, FL, 2012).

10) M. Tegmark, **Our Mathematical Universe** (Knopf, New York, 2014).

11) P. Turchin, Historical Dynamics: Why States Rise and Fall (Princeton University Press; Princeton, NJ; 2003).

CHAPTER 11

Toy Model of the Universe

11.1 INTRODUCTION

The "big bang" theory is the current accepted model of the observable physical universe from its genesis to its present structure. This model describes the evolution of how the universe expanded from an initial state of extreme high temperature and energy density. It value derives from the fact that it provides a mathematical and theoretical framework that explains a broad range of observed phenomena existing in the universe:

- abundance of the light elements (hydrogen, helium, lithium)
- the expansion of the universe
- the existence of the cosmic microwave background
- the large scale structure.

One of the most interesting features of the universe is its composition:

- normal matter – 5%
- dark matter – 27%
- dark energy - 68%

In brief, **normal matter** interacts with the electromagnetic force and under the proper conditions can emit visible light. **Dark matter** does not interact with the electromagnetic force, i.e., it does not absorb, reflect, or emit light. Its existence can only be detected by its gravitational influence on normal matter. Finally, **dark energy** is related to an unknown force that causes the expansion rate of the universe to accelerate over time, rather than slow down as expected from gravitational interactions.

The discipline that investigates these matters is called cosmology. Thus, **cosmology** is the science of the origin and evolution of the physical universe.

Our task in this chapter is to construct a "toy model" of the universe and analyze its properties. We do this to obtain insights into the formation and investigation of the more

DOI: 10.1201/9781003178972-11

complex mathematical models that exist in the research literature. As will be seen shortly, our "toy model" employs rather elementary mathematical techniques. However, many of the essential aspects of the realistic models are embedded in its mathematical structure and conclusions.

The chapter is organized as follows: Section 11.2 introduces several needed ideals and discusses the rules/laws/assumptions required to formulate a whole set of toy universes. Section 11.3 considers some fundamentally uninteresting of "dull" toy universes and their properties. Section 11.4 presents the construction of a nontrivial toy universe and investigates some of its major features. This particular model is called a Fibonacci universe. Since the Fibonacci equation arises in its analysis. Section 11.5 gives a quick introduction to the Fibonacci equation and how it can be solved. This section may be skipped by those who are familiar with this equation. Finally, we present in Section 11.6 a summary of what has been accomplished and comments on how these results can be generalized.

11.2 IN THE BEGINNING: LET THERE BE RULES

A "toy model (TM)" is an elementary model of a system that has many of the inessential details removed and is used to provide insights into the understanding and description of a fuller or more complex model.

A "toy model of the universe (TMOU)" should include the following items and/or features:

(a) There must be an empty or vacuum state.

(b) There must exist particles (matter), that can exist in this universe.

(c) There must exist rules for determing how the particles interact with each other.

(d) There must exist some form of space-time structure in which the TMOU involves.

Our particular class of TMOU is characterized by the following properties:

(i) There are particles which potentially could exist in this universe and they are denoted by

$$\text{“}a\text{”},\ \text{“}b\text{”},\ \text{“}c\text{”},\ldots. \tag{11.2.1}$$

(ii) For a given state of this universe, the particles are arranged in a particular order (from left to right on the in the diagram to follow in the next sections).

(iii) There exist transition rules which govern how one state of the universe evolves to the next state. These rules take the form

$$\begin{cases} a \to f_1(a,b,c,\ldots), \\ b \to f_2(a,b,c,\ldots), \\ \vdots \quad \vdots \end{cases}, \tag{11.2.2}$$

where $(f_1, f_2, \ldots)$ are specified in advance. Note that the transition rules must be internally consistent with each other.

(iv) We may consider the transitions to take place at discrete time intervals, i.e., $t_k = (\Delta t)k$, where $k = (0, 1, 2, 3, \ldots)$ and where the exact value of Δt is not important.

In summary, a TMOU will consider of a priori specified particles and rules for how these particles can interact with each other. However, to construct a given TM, the initial state must be given. Clearly, in general, we expect different initial states to produce different evolved states of the TMOU.

In the next section, we examined several "dull" realizations of these procedures to illustrate the methodlogy.

11.3 SOME "DULL" MODEL UNIVERSES

We examine several TMOU's, all of which are "dull". By "dull," we mean that they do not have interesting or sophisticated dynamics.

Empty Universe

Using the notation

$$\{\ldots\} = \text{contents of the (toy) universe,} \tag{11.3.1}$$

then an empty universe can be indicated by the symbol

$$\{\} = \text{empty university.} \tag{11.3.2}$$

Since there are no particles in this universe, it does not change with (discrete) times, i.e.,

$$\begin{array}{cc} \underline{\text{time}} & \underline{\text{state of universe}} \\ 0 & \{\} \\ 1 & \{\} \\ 2 & \{\} \\ \vdots & \vdots \\ k & \{\} \\ \vdots & \vdots \end{array}, \tag{11.3.3}$$

where the time is expressed in terms of the unit Δt,

Universe with One Particle Type, "a," and $a \to a$

This TMOU contains only particles of one type; they are labeled "a." The transition rule is

$$a \to a, \tag{11.3.4}$$

i.e., if at some discrete time, k, this universe contains N particles of type a, then at the next time, $k+1$, it also contains N particles of type a. To illustrate, let $N = 3$, and obtain

$$\begin{array}{cc} \underline{\text{time}} & \underline{\text{state of universe}} \\ 0 & \{a, a, a\} \\ 1 & \{a, a, a\} \\ 2 & \{a, a, a\} \\ \vdots & \vdots \\ k & \{a, a, a\} \\ \vdots & \vdots \end{array} \quad . \tag{11.3.5}$$

This is an unchanging universe, i.e., given an initial state

$$k = 0, \quad \{\underbrace{a\, a \ldots\, a}_{N}\}, \tag{11.3.6}$$

then at any future (discrete) time, k_1, the universe is in the same state

$$k = k_1, \quad \{\underbrace{a\, a \ldots\, a}_{N}\}. \tag{11.3.7}$$

Universe with One Particle Type, "a," and $a \to aa$

The transition rule for this case is

$$a \to aa. \tag{11.3.8}$$

i.e., each a at time k is replaced by aa at time $k+1$; for example, taking the initial state to be $\{a\ a\ a\}$, then

$$\begin{array}{cc} \underline{\text{time}} & \underline{\text{state}} \text{ of the universe} \\ 0 & \{aaa\} \\ 1 & \{aaaaaa\} \\ 2 & \{aaaaaaaaaaaa\} \\ \vdots & \vdots \\ k & \underbrace{aa \ldots aa}_{3 \cdot 2^k} \\ \vdots & \vdots \end{array} \tag{11.3.9}$$

COMMENT: For the remainder of the chapter, we will drop indications of the discrete time, and also eliminate the brackets for the states of the toy universes. This means that the content of Equation (11.3.8) becomes the expression

$$
\begin{array}{l}
aaa \\
aaaaaa \\
aaaaaaaaaaaa \\
\quad\vdots \\
\underbrace{aa \ldots aa}_{3.2^k} \\
\vdots
\end{array}
\tag{11.3.10}
$$

Universe with Two Particle Types, “a” and “b,” with $a \to a$ and $b \to b$

It should be clear that given an initial state, $I(a, b)$, then under the transition rules

$$a \to a, \quad b \to b, \tag{11.3.11}$$

all future states are also $I(a, b)$, since

$$I(a, b) \text{ at } k \Rightarrow I(a, b) \text{ at } k+1. \tag{11.3.12}$$

Universe with Two Particle Types, “a” and “b,” with $a \to b$ and $b \to a$

For this case, the two particle types switch identities in going from time k to time $k+1$. Three illustrative cases are

$$
\begin{array}{cccc}
a & b & aba & babb \\
b & a & bab & abaa \\
a & b & aba & babb \\
b & a & bab & abaa \\
a & b & aba & babb \\
b & a & bab & abaa \\
\vdots & \vdots & \vdots & \vdots
\end{array}
\tag{11.3.13}
$$

If the initial state is $I_1(a, b)$, then the toy universe oscillates between these two states, i.e.,

$$I_1(a, b) \text{ at } k \Rightarrow I_1(b, a) \text{ at } k+1, \tag{11.3.14}$$

$$I_1(b, a) \text{ at } k+1 \Rightarrow I_1(a, b) \text{ at } k+2, \tag{11.3.15}$$

and the period of the oscillation is 2.

All of the above five toy models are dull or uninteresting because their future evolution can be exactly predicted and no complex structures appear. In the next section, we consider a model that has a number of interesting features that takes it out of the category of being dull or uninteresting.

11.4 NONTRIVIAL TMOU

The model to be constructed in this section is both nontrivial and interesting from the perspective of having rather complex dynamics. It is founded on the following assumptions:

(i) The "matter" of this universe consists of two types of particles: "a" and "b."

(ii) These particles interact such that in going from (discrete) time k to $k + 1$, their transition rules are

$$a \to b, \quad b \to ab. \tag{11.4.1}$$

At this point, several clarifying comments are needed:

(a) The second of the transition rules, $b \to ab$, should be taken to mean that the particle b at time k is replaced by ab in the indicated order "a, followed by b," at the time $k + 1$.

(b) The selection of an initial state is critical for the determination of future states of the TMOU. Different initial states will, in general, produce different evolutions or future states. For our purposes, only simple initial states are selected.

To begin, let the initial state of our TMOU contain only a single particle of type "a." The first eight states evolving from this initial state are

$$\begin{array}{l} a \\ b \\ ab \\ bab \\ abbab \\ bababbab \\ abbabbababbab \\ bababbababbabbababbab \\ abbabbababbabbababbababbabbababbab \\ bababbababbabbababbababbabbabab \\ \qquad\qquad\qquad \ldots babbababbababbabbababbab. \end{array} \tag{11.4.2}$$

In principle, we can continue with this construction to any finite (discrete) time k. However, close inspection of the items in Equation (11.4.2) shows that it contains a lot of structure.

For example:

(1) If the initial state was b rather than a, the evolution would be exactly the same except by a shift in one in the k-values.

(2) Except, for the initial state, the listing has only b's on the right-hand boundary.

(3) The left-handed boundary alternates between a and b.

Note that items (2) and (3) can be proven. First, consider (2). Suppose at time k, the last position is an "a." Then by replacing "a" by "b" we find that the last position is a "b" at time $k+1$. Likewise, if at time k, the last position is a "b," then it gets replaced by "ab" at time $k+1$; thus, giving a "b" at the right boundary. Similarly, at the left boundary if an "a" appears at time k, then at time $k+1$ it is replaced by "b;" if a "b" occurs at time k, it is replaced by "ab" and consequently, an "a" is the particle at the left boundary. Therefore, "a" and "b" will alternate at the left-hand boundary.

Now, what else can be done? We can count the total number of a's and b's at each time k. The results of doing this for the listing in Equation (11.4.2) is given below:

k	$N_k(a)$	$N_k(b)$	$N_k(a+b)$
0	1	0	1
1	0	1	1
2	1	1	2
3	1	2	3
4	2	3	5
5	3	5	8
6	5	8	13
7	8	13	21
8	13	21	34
9	21	34	55

(11.4.3)

where

$$\begin{aligned} N_k(a) &= \text{number of "}a\text{"'s at time } k, \\ N_k(b) &= \text{number of "}b\text{"'s at time } k, \\ N_k(a+b) &= \text{number of "}a\text{"'s and "}b\text{"'s.} \end{aligned}$$

A close inspection of the entries in Equation (11.4.3) shows that for $k \geq z$ (for the limited range of k-values listed above), we have

$$N_{k+2}(\ldots) = N_k(\ldots) + N_{k+1}(\ldots), \tag{11.4.4}$$

where $(\ldots)$ can be a, b or $a+b$. However, Equation (11.4.4) is the Fibonacci equation, a linear difference equation that has been extensively investigated and applied to the modelling of phenomena in many scientific disciplines.

We now show that with the transition rules, stated in Equation (11.4.1), the relationships expressed by Equation (11.4.4) holds for all $k \geq 0$. To do so, begin by denoting an arbitrary initial state by $I(a,b)$, where $I(a,b)$ contains a finite combination of a's and b's.

Let the state at the k-th time level contain. A a's and B b's, where A and B are non-negative integers. Using the transition rules, then the states at levels $k + 1$ and $k + 2$ contain the following numbers of a's and b's:

$$\text{state at } k : A\,(a\text{'}s) + B\,(b\text{'}s) \tag{11.4.5}$$

$$\text{state at } (k+1) : A\,(b\text{'}s) + B\,(a\text{'}s + b\text{'}s) \tag{11.4.6}$$

$$= B\,(a\text{'}s) + (A + B)\,(b\text{'}s)$$

$$\text{state at } (k+2) : B\,(b\text{'}s) + (A + B)\,(a\text{'}s + b\text{'}s) \tag{11.4.7}$$

$$= (A + B)\,(a\text{'}s) + (A + ZB)\,(b\text{'}s)\,.$$

Adding Equations (11.4.5) and (11.4.6), comparing this result to Equation (11.4.7), and, finally, separating these expressions into the "a" and "b" components gives

$$N_{k+2}\,(a) = N_{k+1}\,(a) + N_k\,(a) \tag{11.4.8}$$

$$N_{k+2}\,(b) = N_{k+1}\,(b) + N_k\,(b) \tag{11.4.9}$$

$$N_{k+2}\,(a + b) = N_{k+1}\,(a + b) + N_k\,(a + b)\,. \tag{11.4.10}$$

Continuing with this line of thought, we can ask what is the distribution of the pair of particles labeled "ab." Using the information given in Equation (11.4.2), we find

k	$N_k\,(ab)$
0	0
1	0
2	1
3	1
4	2
5	3
6	5
7	8
8	13

(11.4.11)

where $N_k(ab)$ is the number of "ab" pairs appearing at the k-th time level. This sequence of numbers also satisfies the Fibonacci equation, since from the transition rules, an "a" will only appear in the combination "ab," and this implies that at a gives level, there are as many "ab"'s as there are "a"'s.

The number of "bab" triples also satisfy the Fibonacci equation.

Note that all of the above results start with a single particle, either "a" or "b," initial state. However, all possible combinations of "a"'s and "b"'s may be used as initial states; for example, we have the possibilities:

- $aa,\ bb,\ ab,\ ba$
- $aaa,\ aab,\ aba,\ baa,\ abb,\ bab,\ bba,\ bbb$
- $aaaa,\ aaab,\ aaba,\ abaa,\ baaa,\ aabb,\ abab,\ baab, \ldots$

$$\vdots \qquad \vdots \qquad \vdots$$

- $abbabaaaababbbaababaab$
- etc.

Note, that since the transition rule for b is ordered, i.e., $b \to ab$, and not $b \to ba$, then the evolution of different initial states will give rise to different states at the k-th (discrete) time level. For a simple illustration of this, consider the evolution of the initial state "aa;" we obtain

$$\begin{array}{l} aa \\ bb \\ abab \\ babbab \\ abbababbab \\ bababbabbababbab \\ abbabbababbababbabbababbab \end{array} \tag{11.4.12}$$

and

$$\begin{array}{cccc} \underline{k} & \underline{N_k(a)} & \underline{N_k(b)} & \underline{N_k(ab)} \\ 0 & 2 & 0 & 0 \\ 1 & 0 & 2 & 0 \\ 2 & 2 & 2 & 2 \\ 3 & 2 & 4 & 2 \\ 4 & 4 & 6 & 4 \\ 5 & 6 & 10 & 6 \\ 6 & 10 & 16 & 10 \end{array} \tag{11.4.13}$$

Thus, for the initial state "aa," we have

$$\begin{cases} N_0(a) = 2, & N_1(a) = 0; \\ N_0(b) = 0, & N_1(b) = 2; \\ N_1(ab) = 0, & N_2(ab) = 2; \end{cases} \tag{11.4.14}$$

and $N_k(a)$, $N_k(b)$, and $N_k(ab)$ all satisfy the Fibonacci equation.

A general summary and discussion of this model, the Fibonacci TMOU will be presented in Section 11.6.

11.5 FIBONACCI EQUATION

The Fibonacci equation is a linear, second-order difference equation with constant coefficients. It can be written as

$$F_{k+2} = F_{k+1} + F_k. \tag{11.5.1}$$

Such an equation has two solutions of the form

$$F_k = (\lambda)^k, \tag{11.5.2}$$

where λ is a solution to

$$\lambda^2 - \lambda - 1 = 0. \tag{11.5.3}$$

Solving for λ gives

$$\lambda_1 = \frac{1+\sqrt{5}}{2}, \quad \lambda_2 = \frac{1-\sqrt{5}}{2}, \tag{11.5.4}$$

where

$$\lambda_1 = 1.618\ 033\ 988\ldots, \quad \lambda_2 = -0.618\ 033\ 988\ldots. \tag{11.5.5}$$

Note that

$$\lambda_2 = -\left(\frac{1}{\lambda_1}\right). \tag{11.5.6}$$

The general solution to Equation (11.5.1) is

$$F_k = A(\lambda_1)^k + B(\lambda_2)^k, \tag{11.5.7}$$

where A and B are arbitrary constants. For initial conditions, F_0 and F_1, we have

$$F_0 = A + B, \quad F_1 = A\lambda_1 + B\lambda_2, \tag{11.5.8}$$

which upon solving for A and B gives

$$A = \frac{F_1 - \lambda_2 F_0}{\sqrt{5}}, \quad B = \frac{\lambda_1 F_0 - F_1}{\sqrt{5}}. \tag{11.5.9}$$

Therefore, the general solution to the Fibonacci equation is

$$F_k = \left(\frac{F_1 - \lambda_2 F_0}{\sqrt{5}}\right)(\lambda_1)^k + \left(\frac{F_0\lambda_1 - F_1}{\sqrt{5}}\right)(\lambda_2)^k. \tag{11.5.10}$$

In both the general and research literature the number λ_1 is called the "golden ratio" and represented by the symbol ϕ , i.e.,

$$\phi \equiv \lambda_1 = \frac{1+\sqrt{5}}{2}. \tag{11.5.11}$$

We also have the relation

$$\operatorname*{Lim}_{k\to\infty}\left(\frac{F_{k+1}}{F_k}\right) = \phi. \tag{11.5.12}$$

The standard Fibanocci sequence uses the initial conditions

$$F_0 = 0, \quad F_1 = 1, \tag{11.5.13}$$

and gives

$$F_k : 0, 1, 1, 2, 3, 5, 8, 13, 21, 34, 55, \ldots. \tag{11.5.14}$$

This analysis can be extended to negative values of the index k by using the relation

$$F_{k-2} = F_k - F_{k-1}. \tag{11.5.15}$$

This produces the following results

$$\begin{aligned}
F_{-5} &= 5 \\
F_{-4} &= -3 \\
F_{-3} &= 2 \\
F_{-2} &= -1 \\
F_{-1} &= 1 \\
F_0 &= 0 \\
F_1 &= 1 \\
F_2 &= 1 \\
F_3 &= 2 \\
F_4 &= 3 \\
F_5 &= 5
\end{aligned} \tag{11.5.16}$$

With minor effort, it can be shown that

$$F_{-k} = (-1)^{k+1} F_k. \tag{11.5.17}$$

11.6 DISCUSSION

All of the discussions to come are based on the transition rules

$$a \to b, \quad b \to ab, \tag{11.6.1}$$

for two particle types, "a" and "b," with the initial state being a single "a" particle. In particular, keep the results of Equations (11.4.2) and (11.4.3); and (11.4.8), (11.4.9), and (11.4.10) in mind.

Consideration of what was done in Section (11.4) allows the following conclusions and generalizations to be reached:

(A) Simple rules of transition can produce interesting patterns in the evolution of the possible states of the particles, "a" and "b."

(B) The resulting toy universes are deterministic in the sense that for a given initial state, the patterns or arrangements of the "a"'s and "b"'s are unique.

(C) We may interpret the representations of these toy universes, see Equations (11.4.2) and (11.4.12), as having a pseudo-space-time structure, ie.,

- pseudo-space direction: left to right horizontal direction,
- pseudo-time direction: downward in the vertical direction.

This "universe" is semi-infinite in the horizontal space dimension. It begins at the location of the left most letter (a or b) and extends, unbounded, to the right. If we assign to each particle a space Δx, then we may attribute a location, $X_m = (\Delta x)_m$, if a particle is in the m-th position of the string of "a"'s and "b"'s at a given level.

Likewise, the initial state begins at the discrete time $k = 0$, and transitions to other states by means of the rules presented in Equation (11.6.1). We may introduce a time interval between each neighboring (vertical) level of magnitude Δt. Therefore, the k-th level comes into existence at discrete time $t_k = (\Delta t)\, k$, and interacts with itself to create the new state at level $k + 1$.

(D) Let

$$N_k \equiv N_k(a) + N_k(b), \tag{11.6.2}$$

then with $N_0(a) = 1$, and $N_0(b) = 0$, and $N_1(a) = 0$ and $N_1(b) = 1$, then N_k is given by the expression

$$N_k = \left(\frac{\lambda_1}{\sqrt{5}}\right)(\lambda_1)^k - \left(\frac{\lambda_2}{\sqrt{5}}\right)(\lambda_2)^k, \tag{11.6.3}$$

where

$$\lambda_1 = \frac{1+\sqrt{5}}{2}, \quad \lambda_2 = \frac{1-\sqrt{5}}{2}. \tag{11.6.4}$$

(These results are a consequence of the work presented in Section 11.5 on the solutions of the Fibonacci difference equation.) Since, $\lambda_1 > 1$, it follows that for large k, N_k increases exponentially with k, i.e.,

$$N_k \xrightarrow[\text{large } k]{} \left(\frac{\lambda_1}{\sqrt{5}}\right) e^{\beta k}, \tag{11.6.5}$$

where $\beta = 0.4812118246\ldots$.

If the "size" of this toy universe, R_k, is taken to be

$$R_k \equiv N_k \cdot \Delta x, \tag{11.6.6}$$

then

$$R_k \xrightarrow[\text{large } k]{} (\Delta x)\left(\frac{\lambda_1}{\sqrt{5}}\right) e^{\beta k}. \tag{11.6.7}$$

In other words, this universe is finite for finite k, but is increasing exponentially with k.

(E) Let v_k be the velocity of expansion of the universe at discrete time $t_k = (\Delta t)\, k$, and define it by the relation

$$v_k \equiv \left(\frac{R_k - R_{k-1}}{\Delta t} \right). \tag{11.6.8}$$

Using the definition of R_k, we find

$$v_k = \left(\frac{3}{2\sqrt{5}} \right) \left(\frac{\Delta x}{\Delta t} \right) \left[(\lambda_1)^{k-1} + (\lambda_2)^{k-1} \right], \tag{11.6.9}$$

and this becomes at large k the expression

$$v_k \sim \left(\frac{3}{2\sqrt{5}} \right) \left(\frac{\Delta x}{\Delta t} \right) e^{\beta k}. \tag{11.6.10}$$

(F) For large k, we have

$$V_k \sim HR_k, \tag{11.6.11}$$

where

$$H = \left(\frac{3}{2} \right) \left(\frac{\Delta t}{\lambda_1} \right) = \text{constant}. \tag{11.6.12}$$

This implies that there is a linear relation between the size of the toy universe and its velocity of expansion. A similar formula holds for the actual physical universe and is called the Hubble-Lemaître law.

(G) Finally, it should be observed, again, that the current TMOU have an asymmetry build into their constructions. So, $b \to ab$ and $b \to ba$ are not the same transition rules. To show this consider the initial state of just a single "a" particle, then

$a \to b,\ b \to ab$	$a \to b,\ b \to ba$
a	a
b	b
ab	ba
bab	bab
$abbab$	$babba$
$bababbab$	$babbabab$
$abbabbababbab$	$babbababbabba$

Note that while the total number of particles at each level are the same, they differ in their respective arrangements.

In conclusion, it is remarkable that this elementary TMOU reproduces many of the essential features discovered in the actual physical universe. The work of this chapter can be easily generalized to consider more than two different types of particles; for example, consider three particle types, (a, b, c), with the transition rules

$$a \to b, \quad b \to ac, \quad c \to a, \tag{11.6.13}$$

or

$$a \to b, \quad b \to c, \quad c \to ab. \tag{11.6.14}$$

PROBLEMS

Sections 11.2 and 11.3

1) Consider a TMOU consisting of two types of particles, (a, b). Assume the following transition rules hold

$$a \to ab, \quad b \to a.$$

For the initial state consisting of one particle of type "a," we have

k	S_k
0	a
1	ab
2	aba
3	$abaab$
4	$abaababa$
5	$abaababaabaab$
$\vdots$	$\vdots$

Prove that

$$S_k = S_{k-1}\, S_{k-2}.$$

2) Analyze the dynamics of the three different particle types system, (a, b, c), having the transition rules

$$a \to ab, \quad b \to ac, \quad c \to a,$$

with the initial state "a."

3) For the system

$$a \to b, \quad b \to bc, \quad c \to ab,$$

show that the combination "ba" will never appear in any level if we start with an initial state "a" Derive formulas for the individual number of "a"'s, "b"'s, and "c"'s appearing at the k-th level.

Section 11.4

4) The listing in Equation (11.4.2), based on the rules in Equation (11.4.1) has the following listing for the combination "abb"

k	$N_k(abb)$
0	0
1	0
2	0
3	0
4	1
5	1
6	3
7	4
8	8
⋮	⋮

Is it possible to link N_{k+2} to N_{k+1} and N_k, where $N_k \equiv N_k\ (abb)$?

Section 11.5

5) Prove the following relations

- $\sum_{k=1}^{N} F_k = F_{N+2} - 1$
- $\sum_{k=1}^{N} F_k^2 = F_N F_{N+1}$
- $\sum_{k=0}^{\infty} F_k x^k = \dfrac{x}{1 - x - x^2}$.

NOTES AND REFERENCES

Section 11.1: The following books provide good introductions to the structure of the universe

(1) S. Weinburg, **The First Three Minutes: A Modern View of the Origin of the Universe** (Basic Books, New York, 1993).

(2) M.J. Rees, **New Perspectives in Astro-Physical Cosmology** (Cambridge University Press, Cambridge, 2000).

(3) A. Liddle, **An Introduction to Modern Cosmology** (Wiley, New York, 2015).

(4) M. Roos, **Introduction to Cosmology 4th Edition** (Wiley, New York, 2015).

The book by Liddle gives in Chapter 1 an excellent, concise history of cosmological concepts and models. Also, it will be of value to read, as quick introductions, the WIKIPEDIA web items on the subjects

- "Big Bang"
- Chronology of the Universe

- "Cosmology"
- "Observable Universe"
- Universe.

Sections 11.2, 11.3, and 11.4: While the models that we construct are elementary and the level of mathematical knowledge required to analyze them is not advanced, there is a rather larger literature on the underlying mathematical basis. For the details, see the following publications

(5) M. Barge and J. Kwapisz, Geometric theory of unimodular pisot substitutions, **American Journal of Mathematics**, Vol. 128, No. 5 (2006), 1219-1282.

(6) P. Arnoux and E. Harriss, What is a Rauzy fractal? **Notices of the American Mathematics Society**, Vol. 61, Number 7 (2014), 768-770.

(7) G. Rauzy, Nombres algébriques et substitutions, **Bulletin de la Societé Mathématiqu de France**, Vol. 110 (1982), 147-178.

Section 11.5: Fibonacci numbers and their relationship to other "named" numbers and sequences are discussed in the volume

(8) R.A. Dunlap, **The Golden Ratio and Fibonacci Numbers** (World Scientific, London, 1997).

Applications to several areas of the natural sciences are also presented.

CHAPTER 12

Diffusion and Heat Equations

12.1 INTRODUCTION

The partial differential equation

$$\frac{\partial u(x,t)}{\partial t} = D\frac{\partial^2 u(x,t)}{\partial x^2}, \tag{12.1.1}$$

where D is a constant appears in many physical science and mathematical contexts, and is called either the diffusion or heat equation. It is a special case of the parabolic partial differential equation

$$u_t = F(x,t,u_x,u_{xx}), \tag{12.1.2}$$

where the following standard notation is used

$$u_t \equiv \frac{\partial u}{\partial t}, \quad u_x \equiv \frac{\partial u}{\partial x}, \quad u_{xx} \equiv \frac{\partial^2 u}{\partial x^2}. \tag{12.1.3}$$

Observe that Equation (12.1.1) is invariant under the transformation

$$x \to -x. \tag{12.1.4}$$

This means that for $-\infty < x < \infty$, then if $u(x,t)$ is a solution, then $u(-x,t)$ is also a solution. However, if $t \to -t$, the equation is not invariant. Physically, these results have the interpretations:

(i) For, $-\infty < x < \infty$, the system does not distinguish left from right. So a system of particles concentrated near the origin at $t = 0$, can spread or diffuse both in the left and right directions.

(ii) For $t \to -t = \bar{t}$, we have

$$u(x,t) \to u(x,-t) = v(x,\bar{t}) \tag{12.1.5}$$

and

$$\frac{\partial v(x,\bar{t})}{\partial \bar{t}} = -D\frac{\partial^2 v(x,\bar{t})}{\partial x^2}. \tag{12.1.6}$$

DOI: 10.1201/9781003178972-12

Thus, while the right side of Equation (12.1.1) has a positive sign, the corresponding sign in this last equation is negative. This means that the equation for $v(x, \bar{t})$ models a system, where with increase in $\bar{t}$, the particle concentration increases, i.e., anti-diffusion takes place.

As hinted at above, Equation (12.1.1) models both diffusion and heat problem. It can be extended to include D depending on (x, t, u) and the addition of other terms modelling reaction, diffusion, and advection. Several well known equations are:

Linear advection – diffusion-reaction equation

$$u_t + a_1 u_x = D u_{xx} + \lambda_1 u \tag{12.1.7}$$

Burgers' equation

$$u_t + a_2 u u_x = D u_{xx} \tag{12.1.8}$$

Fisher equation

$$u_t = D u_{xx} + \lambda_2 u (1 - u) \tag{12.1.9}$$

Nonlinear diffusion equation

$$u_t = D (u u_x) x. \tag{12.1.10}$$

In these partial differential equations, the coefficients, $(a_1, a_2, D, \lambda_1, \lambda_2)$ are constant parameters. Note that Equations (12.1.7) and (12.1.8) have, respectively, linear and nonlinear advection; Equations (12.1.7) and (12.1.9) have, respectively, linear and nonlinear reaction terms; and Equation (12.1.10) has a diffusion coefficient that is linear in u, thus making the equation a nonlinear diffusion equation.

Using the indicated notation given in Equation (12.1.3), Equation (12.1.1) becomes

$$u_t = D u_{xx}, \qquad u = u(x, t). \tag{12.1.11}$$

The purposes of this chapter can be stated as follows:

(I) derive the heat equation using physical arguments relating to temperature differences in a long slinder rod;

(II) derive the diffusion equation from the consideration of a random walk;

(III) obtain a second derivation of the diffusion equation using probability theory;

(IV) solve the physical problem of a heated rod with given boundary and initial values.

These several ways of obtaining the diffusion and heat differential equations also show the connections between their "discrete" origins and the resulting "continuous" partial differential equations.

The order of the chapter follows the listing of the above items, (I) to (IV). However, we also include, in Section 12.5, a brief discussion on several difficulties which arise in going from the discrete to the continuous limits of diffusion/heat phenomena.

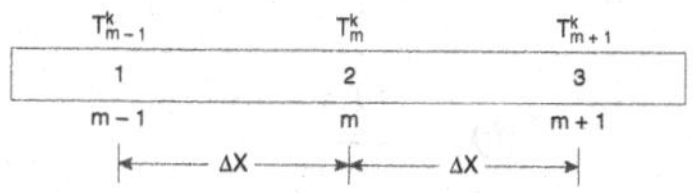

T^k_{m-1} T^k_m T^k_{m+1}

1 2 3

m − 1 m m + 1

ΔX ΔX

Figure 12.1: Temperature distribution in a long metal rod at locations (x_{m-1}, x_m, x_{m+1}), at t_k.

12.2 HEAT EQUATION DERIVATION

We will "derive" or construct the heat partial differential equation

$$u_t = Du_{xx}, \qquad u = u(x,t), \tag{12.2.1}$$

from consideration of a simple physical system, namely, a long, thin metal rod aligned along the x-axis. Let $T(x,t)$ be the temperature at position x at the time t. To begin, we focus our attention on the temperature at only a discrete set of points, separated by the same distance Δx. Further, we only observe the system at a discrete set of times, separated by the interval Δt. Therefore,

$$\begin{cases} t \to t_k = (\Delta t)\, k, & x \to x_m = (\Delta x)\, m, \\ T(x,t) \to T^k_m, & \\ k = (0, 1, 2, \ldots), & m = (\text{integers}). \end{cases} \tag{12.2.2}$$

The following assumptions are made (see Figure 12.1):

(a) T^k_m is the temperature at the location x_m, at the discrete time t_k.

(b) The temperature at a given discrete point is only determined by its values at the nearest neighboring points, i.e., the temperature at x_m is related to the temperatures at x_{m-1} and X_{m+1}.

(c) The temperature change, in going from t_k to t_{k+1} at a given point, is proportional to the temperature differences between this point and its nearest neighbors.

Expressed mathematically, these assumptions take the form

$$\begin{aligned} T^{k+1}_m - T^k_m &= D_1\left(T^k_{m-1} - T^k_m\right) + D_1\left(T^k_{m+1} - T^k_m\right) \\ &= D_1\left(T^k_{m+1} - 2T^k_m + T^k_{m-1}\right), \end{aligned} \tag{12.2.3}$$

where D_1 is the constant of proportionally. Note that D_1 will depend on the specific properties of the metal and we assume them to be such that D_1 may be taken as constant for this construction.

Dividing the left side of Equation (12.2.3) by Δt, dividing the right side by $(\Delta x)^2$, and putting in the necessary factors of Δx and Δt so that the overall resulting expression is algebraically correct, gives

$$\frac{T^{k+1}_m - T^k_m}{\Delta t} = D\left[\frac{T^k_{m+1} - 2T^k_m + T^k_{m-1}}{(\Delta x)^2}\right], \tag{12.2.4}$$

where

$$D = \frac{D_1(\Delta x)^2}{\Delta t}. \tag{12.2.5}$$

Since

$$\lim_{\Delta t \to 0}\left[\frac{y(t+\Delta t) - y(t)}{\Delta t}\right] = \frac{dy(t)}{dt}, \tag{12.2.6}$$

$$\lim_{\Delta x \to 0}\left[\frac{y(x+\Delta x) - 2y(x) + y(x-\Delta x)}{(\Delta x)^2}\right] = \frac{d^2 y(x)}{dx^2}, \tag{12.2.7}$$

then taking the limits, $\Delta t \to 0$ and $\Delta x \to 0$, in Equation (12.2.4) such that

$$\frac{(\Delta x)^2}{\Delta t} = \text{constant}, \tag{12.2.8}$$

produces the result

$$\frac{\partial T(x,t)}{\partial t} = D\frac{\partial^2 T(x,t)}{\partial x^2}, \tag{12.2.9}$$

and this is just Equation (12.2.1). The parameter D is the "heat coefficient" for the metal.

The discretization, i.e., $T(x,t) \to T_m^k$, given in Equation (12.2.4), provides a finite difference scheme for the numerical integration of the heat partial differential equation. To see this, let R be defined as

$$R = \frac{\Delta t}{(\Delta x)^2}, \tag{12.2.10}$$

and rewrite Equation (12.2.4) to the form

$$T_m^{k+1} = DR\left(T_{m+1}^k + T_{m-1}^k\right) + (1 - 2DR)\,T_m^k. \tag{12.2.11}$$

Since T_m^k, k fixed and all relevant m, must be non-negative (why?), it follows that a way to this condition holds is to make the requirement

$$1 - 2DR \geq 0 \Rightarrow \Delta t \leq \frac{(\Delta x)^2}{2D}. \tag{12.2.12}$$

Consider the situation where a rod of length L has its temperatures fixed at the left end at T_0 and at T_m at the right end. Eventually a steady state or equilibrium temperature distribution will occur. The case corresponds to $T_t(x,t) = 0$ or

$$T_{xx} = 0 \Rightarrow T(x,\infty) = c_1 + c_2 x, \tag{12.2.13}$$

with

$$T(0,\infty) = T_0, \qquad T(L,\infty) = T_m. \tag{12.2.14}$$

Therefore,

$$T_0 = c_1, \quad T_m = c_1 + c_2 L, \tag{12.2.15}$$

and

$$c_1 = T_0, \quad c_2 = \frac{T_m - T_0}{L}, \tag{12.2.16}$$

which gives

$$\begin{cases} T(x,\infty) = T_0 + \left(\frac{T_m - T_0}{L}\right)x, \\ 0 \le x \le L. \end{cases} \tag{12.2.17}$$

For the discrete case of this equilibrium problem, we have ($T_m^{\infty} \equiv T_m$)

$$T_{m+1} - 2T_m + T_{m-1} = 0, \tag{12.2.18}$$

which has the solution

$$T_m = A + B_m, \tag{12.2.19}$$

with boundary conditions

$$T_0 = \text{given}, \quad T_m = BM, \tag{12.2.20}$$

where M on the right side of T_m is now discrete and has the integer value

$$M = \frac{L}{\Delta x}. \tag{12.2.21}$$

Solving for A and B, and placing these values into Equation (12.2.19) gives

$$T_m = T_0 + \left(\frac{T_m - T_0}{L}\right)x_m, \tag{12.2.22}$$

which is just a discretized version of Equation (12.2.17).

12.3 DIFFUSION EQUATION AND RANDOM WALKS

Consider a random walk in one space dimension, for example, along the x-axis where discrete locations are indicated by $x_m = (\Delta x)\,m$. A Particle at x_m at discrete time, $t_k = (\Delta t)\,k$, can move to the right with probability p and thus can also move to the left with probability q where

$$p + q = 1. \tag{12.3.1}$$

Let u_m^k be the probability to find the particle at position x_m at time t_k. This probability satisfies the following equation

$$u_m^{k+1} = pu_{m+1}^k + qu_{m-1}^k, \tag{12.3.2}$$

i.e., the probability to find the particle at x_m at t_{k+1} is the sum of the probability for it to be at x_{m-1} at time t_k, times the probability to move to the right, and the probability to be at x_{m+1} at time t_k, times the probability to move to the left. See Figure 12.2.

Subtract u_m^k from both sides of Equation (12.3.2) and then rearrange terms to get the expression

$$\begin{aligned} \left(u_m^{k+1} - u_m^k\right) &= p\left(u_{m+1}^k - 2u_m^k + u_{m-1}^k\right) \\ &\quad + (2p-1)\left(u_m^k - u_{m-1}^k\right). \end{aligned} \tag{12.3.3}$$

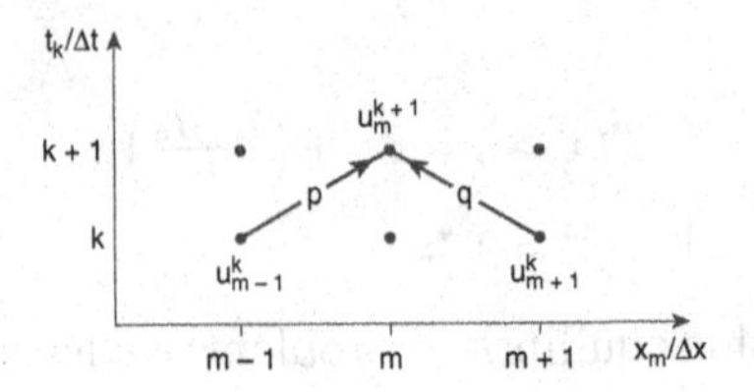

Figure 12.2: Possible movements to get to the point (x_m, t_{k+1}).

In turn, this equation can be rewritten as follows

$$\frac{u_m^{k+1} - u_m^k}{\Delta t} = \left(\frac{p}{\Delta t}(\Delta x)^2\right)\left[\frac{u_{m+1}^k - 2u_m^k + u_{m-1}^k}{(\Delta x)^2}\right] + \frac{2p-1}{\Delta t}(\Delta x)\left[\frac{u_m^k - u_{m-1}^k}{\Delta x}\right]. \tag{12.3.4}$$

If we take the proper limits, then

$$\lim_{\Delta t \to 0}\left(\frac{u_m^{k+1} - u_m^k}{\Delta t}\right) = \frac{\partial u(x,t)}{\partial t} \equiv u_t(x,t), \tag{12.3.5}$$

$$\lim_{\Delta x \to 0}\left[\frac{u_{m+1}^k - 2u_m^k + u_{m-1}^k}{(\Delta x)^2}\right] = \frac{\partial^2 u(x,t)}{\partial x^2} \equiv u_{xx}(x,t). \tag{12.3.6}$$

However, we must be careful with the two coefficients on the right side of Equation (12.3.4). For example, the coefficient of the first term forces us to take

$$\left(\frac{p}{\Delta t}\right)(\Delta x)^2 = D = \text{constant}. \tag{12.3.7}$$

This means that a proper differential equation will be obtained only if Δx and Δt go to zero (simultaneously) such that

$$\Delta t \propto (\Delta x)^2, \tag{12.3.8}$$

Further, if we require

$$p = \frac{1}{2} + \left(\frac{b}{2}\right)\Delta x + O\left[(\Delta x)^2\right], \tag{12.3.9}$$

$$q = \frac{1}{2} - \left(\frac{b}{2}\right)\Delta x + O\left[(\Delta x)^2\right], \tag{12.3.10}$$

then it follows that

$$(2p-1)\left(\frac{\Delta x}{\Delta t}\right) = \left\{b\Delta x + O\left[(\Delta x)^2\right]\right\}\left(\frac{\Delta x}{\Delta t}\right) = b\left[\frac{(\Delta x)^2}{\Delta t}\right] + O\left\{(\Delta x)\left[\frac{(\Delta x)^2}{\Delta t}\right]\right\}, \tag{12.3.11}$$

and taking the limits gives

$$\lim_{\substack{\Delta t \to 0 \\ \Delta x \to 0 \\ \frac{(\Delta x)^2}{\Delta t} = \text{fixed}}} (2p-1)\left(\frac{\Delta x}{\Delta t}\right) = \bar{v}, \tag{12.3.12}$$

where

$$\bar{v} = bR, \quad R = \frac{(\Delta x)^2}{\Delta t}. \tag{12.3.13}$$

Thus, the "discretization" presented in the above Equation (12.3.2) corresponds in the continuum limit to the partial differential equation

$$u_t = Du_{xx} + \bar{v}u_x. \tag{12.3.14}$$

This is a linear advection-diffusion equation with advection velocity, $-\bar{v}$, and diffusion coefficient D. Note that this result is for the situation where $p \neq q$. However, if $p = q$, then the standard diffusion equation is obtained i.e.,

$$p = q \Rightarrow \bar{v} = 0 \rightarrow u_t = Du_{xx}. \tag{12.3.15}$$

12.4 DIFFUSION AND PROBABILITY

This section is essentially another version of what was demonstrated in Section 12.3. We closely follow the arguments presented by Murray (1989, pps. 232-236).

Assume that a particle moves on the x-axis such that it can only occupy the discretized locations $x_m = (\Delta x)m$ where Δx is a fixed length and m is an integer. Further assume that this particle is initially at $x = 0$, at $t = 0$, and moves randomly at each interval Δt of time, either to the left or right with probability one-half. The probability of the particle being at $x_m = (\Delta x)m$ at time $t_k = (\Delta t)k$ will be denoted as P_m^k. The task is to calculate P_m^k and then use this relation to construct an asymptotic continuum expression in the variables x and t, where

$$\begin{cases} \Delta x \rightarrow 0, \quad \Delta t \rightarrow 0, \\ m \rightarrow \infty, \quad k \rightarrow \infty, \\ x = (\Delta x)m = \text{fixed}, \\ t = (\Delta x)k = \text{fixed}. \end{cases} \tag{12.4.1}$$

To continue, we define the integer valued numbers, a and b, to be

a = number of steps that the particle takes to the right,

b = number of steps taken by the particle to the left.

Therefore,

$$m = a - b, \qquad k = a + b. \tag{12.4.2}$$

and from the meaning of the probability P_m^k, it follows that

$$p_m^k \equiv \frac{(\text{number of k-step paths leading to the location } x_m)}{(\text{total number of k-step paths})}. \tag{12.4.3}$$

Note that there are 2^k possible k-step paths. Further, the number of k-step paths leading to the location x_m is

$$C_a^k = \frac{k!}{a!\,(k-a)!} = \frac{k!}{a!b!}, \tag{12.4.4}$$

where C_a^k is the binomial coefficient defined by the relation

$$(x+y)^k = \sum_{a=0}^{k} C_a^k x^{k-a} y^a. \tag{12.4.5}$$

Therefore, P_m^k is

$$P_m^k = \frac{k!}{2^k \left(\frac{k+m}{2}\right)! \left(\frac{k-m}{2}\right)!}. \tag{12.4.6}$$

Likewise, we expect

$$\sum_{m=-k}^{k} P_m^k = 1, \tag{12.4.7}$$

since the sum of the probabilities must be equal to 1. This can be easily shown by using the result in Equation (12.4.5) with $x = y = \frac{1}{2}$, i.e.,

$$\begin{aligned} \sum_{m=-k}^{k} P_m^k &= \sum_{a=0}^{k} C_a^k \left(\frac{1}{2}\right)^{k-a} \left(\frac{1}{2}\right)^a \\ &= \left(\frac{1}{2} + \frac{1}{2}\right)^k \\ &= 1. \end{aligned} \tag{12.4.8}$$

We now turn to a continuous approximation to P_m^k. To begin, we need the so called stirling formula for the factorial expressions, i.e., for large n

$$n! \sim (2\pi n)^{\frac{1}{2}} n^n e^{-n}, n \text{ large}. \tag{12.4.9}$$

If this result is used in Equation (12.4.6), under the assumptions that

$$m \gg 1, \quad k \gg 1, \quad k - m \gg 1, \tag{12.4.10}$$

the

$$P_m^k \sim \left(\frac{2}{\pi k}\right)^{\frac{1}{2}} \exp\left[(-1)\left(\frac{m^2}{2k}\right)\right]. \tag{12.4.11}$$

If we now make the replacements

$$(\Delta x)\, m \to x, \quad (\Delta t)\, k \to t, \tag{12.4.12}$$

and divide Equation (12.4.11) by $2\Delta x$, then we obtain

$$\frac{P\left(\frac{x}{\Delta x}, \frac{t}{\Delta t}\right)}{2\Delta x} \sim \left\{\frac{1}{4\pi t\left[\frac{(\Delta x)^2}{2\Delta t}\right]}\right\}^{\frac{1}{2}} . \exp\left\{\frac{(-)\,x^2}{4t\left[\frac{(\Delta x)^2}{2\Delta t}\right]}\right\}, \tag{12.4.13}$$

where

$$P\left(\frac{x}{\Delta x}, \frac{t}{\Delta t}\right) = P_m^k. \tag{12.4.14}$$

Let us now take the limits $\Delta x \to 0$ and $\Delta t \to 0$, such that the following condition holds

$$\underset{\substack{\Delta t \to 0 \\ \Delta x \to 0}}{\text{Lim}} \left[\frac{(\Delta x)^2}{2\Delta t}\right] = D = \text{constant}. \tag{12.4.15}$$

If we now define

$$u(x, t) \equiv \underset{\substack{\Delta t \to 0 \\ \Delta x \to 0}}{\text{Lim}} \frac{\left(P\frac{x}{\Delta x}, \frac{t}{\Delta t}\right)}{2\Delta x}, \tag{12.4.16}$$

then

$$u(x, t) = \left(\frac{1}{4\pi D t}\right)^{\frac{1}{2}} \exp\left[(-)\left(\frac{x^2}{4Dt}\right)\right]. \tag{12.4.17}$$

Note that the left side of the relation in Equation (12.4.13) is the probability density (probability per length) of locating the particle in the interval $(x - \Delta x)$ and $(x + \Delta x)$.

Consequently, $u\,(x, t)$ is also a probability density associated with a particle initially released at $x = 0$, moving randomly at each value of t, to the left or right with probability 0.5 for each movement.

We now demonstrate that the function $u\,(x, t)$ is a solution to the diffusion differential equation. This will be done by calculating $u_t\,(x, t)$ and $u_{xx}\,(x, t)$, and observing that they are equal.

Comment

For our purposes, we will set $D = 1$ and, as a consequence, the partial differential equation of interest is

$$v_t = v_{xx}. \tag{12.4.18}$$

Our task is to show that a solution is $v = u$, where u is the expression given in Equation (12.4.17).

First, let us calculate u_t. We have

$$\begin{aligned}\frac{\partial u\,(x, t)}{\partial t} &= \left(\frac{1}{\sqrt{4\pi}}\right)\left(-\frac{1}{2}\right) t^{-3/2} \exp\left[-\left(\frac{x^2}{4t}\right)\right] \\ &\quad + \left\{\left(\frac{1}{\sqrt{4\pi t}}\right) \exp\left[-\left(\frac{x^2}{4t}\right)\right]\right\} \frac{d}{dt}\left(-\frac{x^2}{4t}\right) \\ &= \left\{\frac{(-1)\exp\left[-\left(\frac{x^2}{4t}\right)\right]}{\sqrt{16\pi t^3}}\right\}\left(1 + \frac{x^2}{2t}\right).\end{aligned} \tag{12.4.19}$$

Likewise, calculating u_x and u_{xx}, respectively, give

$$\frac{\partial u(x,t)}{\partial x} = \left[\frac{(-1)}{\sqrt{16\pi t^3}}\right] x \exp\left[-\left(\frac{x^2}{4t}\right)\right], \tag{12.4.20}$$

and

$$\frac{\partial^2 u(x,t)}{\partial x^2} = \left\{\frac{(-1)\exp\left[-\left(\frac{x^2}{4t}\right)\right]}{\sqrt{16\pi t^3}}\right\}\left(1+\frac{x^2}{2t}\right). \tag{12.4.21}$$

Therefore, except for $t = 0$, we have

$$u_t(x,t) = u_{xx}(x,t). \tag{12.4.22}$$

Note that if $t \rightarrow Dt$, then the expression in Equation (12.4.17) is reproduced.

If $u(u,t)$ is a probability density, then the following condition holds,

$$\frac{d}{dt}\int_{-\infty}^{\infty} u(x,t)\,dx = 0, \tag{12.4.23}$$

with

$$\int_{-\infty}^{\infty} u(x,t)\,dx = 1. \tag{12.4.24}$$

To prove the first relation start with

$$\frac{\partial u}{\partial t} = \frac{\partial^2 u}{\partial x^2}, \tag{12.4.25}$$

and integrate both sides from $(-\infty)$ to $(+\infty)$. Doing this gives

$$\int_{-\infty}^{\infty} \frac{\partial u}{\partial t}\,dx = \int_{-\infty}^{\infty} \frac{\partial^2 u}{\partial x^2}\,dx, \tag{12.4.26}$$

and, consequently,

$$\begin{aligned}\frac{d}{dt}\int_{-\infty}^{\infty} u(x,t)\,dx &= \int_{-\infty}^{\infty} \frac{\partial}{\partial x}\left(\frac{\partial u}{\partial x}\right) dx \\ &= \int_{-\infty}^{\infty} d\left(\frac{\partial u}{\partial x}\right) \\ &= \left(\frac{\partial u}{\partial x}\right)\bigg|_{x=\infty} - \left(\frac{\partial u}{\partial x}\right)\bigg|_{x=-\infty} \\ &= 0 - 0 = 0,\end{aligned} \tag{12.4.27}$$

where it is assumed that

$$\lim_{|x|\to\infty} \frac{\partial u}{\partial x} = 0. \tag{12.4.28}$$

The above arguments give the result of Equation (12.4.23).

Next, consider the integral

$$\begin{aligned}
\int_{-\infty}^{\infty} u(x,t)\,dx &= \left(\frac{1}{\sqrt{4\pi Dt}}\right) \int_{-\infty}^{\infty} \exp\left(-\frac{x^2}{4Dt}\right) dx \\
&= \left(\frac{1}{\sqrt{\pi}}\right) \int_{-\infty}^{\infty} \exp\left(-y^2\right) dy \\
&= 1,
\end{aligned} \tag{12.4.29}$$

where in the second relation on the right side, the following change of variable was made

$$y = \frac{x}{\sqrt{4Dt}}. \tag{12.4.30}$$

Thus, this calculation proves Equation (12.4.24).

The solution, $u(x, t)$, given in Equation (12.4.17) is called the **fundamental solution** of the heat or diffusion equation. If it is denoted by

$$\Phi(x, t) \equiv \left(\frac{1}{\sqrt{4\pi Dt}}\right) \exp\left[-\left(\frac{x^2}{4Dt}\right)\right], \tag{12.4.31}$$

then $\Phi(x, t)$ has the following properties:

(i) $\Phi(x, t) > 0$, for $-\infty < x < \infty, \quad t > 0$.

(ii) For $t > 0$, $\Phi(x, t)$ has derivatives of all orders in both x and t.

(iii) $\Phi(x, t)$, for $t > 0$, has an area equal to 1, i.e.,

$$\int_{-\infty}^{\infty} \Phi(x, t)\, dx = 1.$$

From (iii), we conclude that since the area under the graph of $\Phi(x, t)$ vs x, for any value of $t > 0$, is 1, then the maximum of $\Phi(x, t)$ goes to zero as $t \to \infty$. This means that the tails of $\Phi(x, t)$ spread out more and more as t becomes larger. See Figure 12.3.

We conclude with the following additional observations:

(a) The heat/diffusion equation is invariant under transitions in x and t, i.e.,

$$x \to \bar{x} + x_0, \quad t \to \bar{t} + t_0, \tag{12.4.32}$$

where x_0 and t_0 are constant. This means that

$$u_t = Du_{xx} \to u_{\bar{t}} = Du_{\bar{x}\bar{x}}. \tag{12.4.33}$$

Consequently, if $u(x, t)$ is a solution, then $u(\bar{x} + x_0\,,\ \bar{t} + t_0)$ is also a solution.

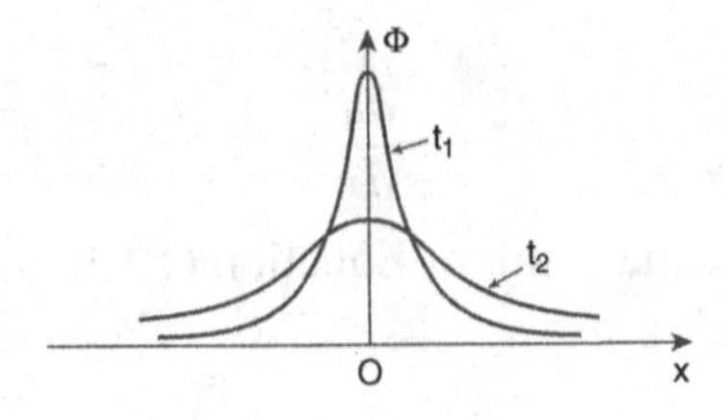

Figure 12.3: Plot of $\Phi(x, t)$ vs x for two times, $t_2 > t_1$.

(b) Let $\lambda > 0$ be a constant. The scaling transformation

$$x \to \lambda x, \quad t \to \lambda^2 t, \tag{12.4.34}$$

leaves the heat/diffusion equation invariant. This means that for the change of variables

$$x_1 = \lambda x, \quad t_1 = \lambda^2 t, \tag{12.4.35}$$

the partial differential equation retains the same structure,

$$u_{t_1} = D u_{x_1 x_1}. \tag{12.4.36}$$

(c) If we consider the **Cauchy problem** for the heat/diffusion equation, i.e.,

$$\begin{cases} u_t = D\, u_{xx}, \ |x| < \infty, \ t > 0, \\ u(x, 0) = g(x), |x| < \infty, \end{cases} \tag{12.4.37}$$

where $g(x)$ is specified, then the solution is

$$u(x, t) = \int_{-\infty}^{\infty} \Phi(x - y, t)\ g(y)\ dy. \tag{12.4.38}$$

See Logan (1994, pps. 42-45).

12.5 DERIVATION DIFFICULTIES

Three separate arguments have been presented for the derivation of the heat/diffusion partial differential equation

$$u_t = D u_{xx}, \quad u = u(x, t). \tag{12.5.1}$$

These arguments involved heat (really temperature) distribution, random walks, and probability. However, the above heat/diffusion equation can not be fundamental because it violates the special theory of relativity in the sense that heat/diffusion phenomena, as modelled by this equation, propagates with an infinite speed. One technique to resolve this issue is to generalize the standard heat/diffusion equation to the form

$$\tau u_{tt} + u_t = D u_{xx}, \tag{12.5.2}$$

where the positive parameter τ is called the relaxation time. Note that Equation (12.5.2) is a damped wave equation with a finite speed of propagation c, given by

$$c = \left(\frac{D}{\tau}\right)^{\frac{1}{2}}. \tag{12.5.3}$$

Observe that when $\tau \to 0$, Equation (12.5.2) goes over to Equation (12.5.1) and c becomes infinite, i.e.,

$$\operatorname*{Lim}_{\tau \to 0} c = \infty. \tag{12.5.4}$$

The Equation (12.5.2) is called the relativistic heat equation. But, it also has several problems associated with its mathematical structure:

(i) The heat/diffusion equation is a parabolic partial differential equation and requires a knowledge of $u(x, 0)$, plus boundary conditions.

(ii) The realistic heat equation is a hyperbolic equation and requires information on $u_t(x, 0)$ to obtain a solution.

(iii) The heat/diffusion equation's physical solutions require a positivity condition to be satisfied, i.e.,

$$u_t = Du_{xx} : u(x, 0) \geq 0 \Rightarrow u(x, t) \geq 0, \ t > 0. \tag{12.5.5}$$

Since Equation (12.5.2) is a hyperbolic equation, its solutions want to "wave" and this makes it difficult to maintain positivity.

Both the standard heat/diffusion and the relativistic heat equations are linear partial differential equations. An ansatz, providing a nonlinear generalization, is the following expression

$$u_t = D\frac{\partial}{\partial x}\left[\frac{uu_x}{\sqrt{u^2 + \left(\frac{D^2}{c^2}\right)(u_x)^2}}\right], \tag{12.5.6}$$

where c is the speed of light. Note that when $c \to \infty$, then this equation reduces to the usual heat/diffusion equation. Also, observe that if

$$\left(\frac{D}{c}\right)|u_x| \ll u, \tag{12.5.7}$$

then Equation (12.5.6) is replaced by the approximate, very complicated equation

$$u_t = D\left[1 - \left(\frac{D^2}{2c^2}\right)\left(\frac{u_x}{u^2}\right)\right]u_{xx} + \left(\frac{D^3}{2c^2}\right)\left(\frac{u_x}{u}\right)^3. \tag{12.5.8}$$

Our major conclusion is to not take Equations (12.5.6) or (12.5.8) seriously as mathematical models for application to realistic physical systems.

The constructions or derivations of the heat/diffusion equations have essentially been based on physical models of phenomena taking place at the microscopic level, with particles, (atoms) located at discrete positions, with hops or displacements from one position to neighboring, positions occurring at discrete time intervals. Clearly, actual physical systems do not have this particular behavior. Also, the appearance of Δx's and Δt's gives additional reasons not to fully believe in what we have accomplished. In the end, we marvel at all the excellent (and accurate) results that have been achieved when these equations are applied to large classes of systems in the physical and engineering systems.

Finally, it would be an interesting and important task to try to relate the Δx's and Δt's back to physical quantities such as interatomic distances, relaxation times, and other physical parameters. Mathematically, all of these issues are connected to the positivity property of the solutions and the need to have finite propagation speeds.

12.6 HEATED ROD PROBLEM

Consider a long, thin rod, of length L, insulated except for the two ends located at $x = 0$ and $x = L$. Let $u(x, t)$ be the temperature at a point x, at time t. Further, assume that the temperature at $x = 0$ and $x = L$ are held fixed, and the initial temperature distribution $u(x, 0) = f(x)$ is given. If the material properties of the rod are taken to be constant along the length of the rod, then the temperature $u(x, t)$ will satisfy the heat equation

$$u_t(x, t) = Du_{xx}(x, t), \tag{12.6.1}$$

where

$$D = \frac{\kappa}{s\rho} = \text{constant}, \tag{12.6.2}$$

and

κ = termal conductivity of the rod material

s = specific heat of the rod material

ρ = mass density of the rod.

The task is to obtain from $u(x, 0) = f(x)$, and the initial/boundary values, the temperature $u(x, t)$ for $t > 0$. Mathematically expressed, we have

$$u_t(x, t) = Du_{xx}(x, t) \tag{12.6.3}$$

$$u(0, t) = u_1, \quad u(L, t) = u_2, \tag{12.6.4}$$

$$u(x, 0) = f(x), \tag{12.6.5}$$

with the physical requirements

$$f(0) = u_1, \quad f(L) = u_2. \tag{12.6.6}$$

These restrictions are to hold for

$$0 \le x \le L, \quad t > 0. \tag{12.6.7}$$

To obtain the required solution, we will use the **method of separation of variables**, i.e., represent $u(x, t)$ as

$$u(x, t) = X(x)\, T(t). \tag{12.6.8}$$

Therefore,

$$u_t = XT', \quad u_{xx} = X''T, \tag{12.6.9}$$

where

$$X'' = X''(x) = \frac{d^2X(x)}{dx^2}, \quad T' = T'(t) = \frac{dT(t)}{dt}, \tag{12.6.10}$$

and

$$X\,T' = DX''T. \tag{12.6.11}$$

If this equation is divided by DXT, then

$$\frac{T'(t)}{DT(t)} = \frac{X''(x)}{X(x)} = -\lambda, \tag{12.6.12}$$

where λ is a constant. The reason why λ must be a constant is because the left side of this equation depents only on t, while the right side is a function of x. Since x and t are independent variables, this relationship can be true only if each ratio of functions is a constant.

For the remainder of this section, we will take u_1 and u_2 to be zero, i.e.,

$$u_1 = 0, \quad u_2 = 0. \tag{12.6.13}$$

Also, since we know (physically) that, under the conditions of Equation (12.6.13), we must have

$$\lim_{t\to\infty} u(x, t) = 0, \tag{12.6.14}$$

it follows that λ must be positive. Thus, $T(t)$ and $X(x)$ satisfy the ordinary differential equations

$$T'(t) = -\lambda DT(t), \tag{12.6.15}$$

$$X''(x) = -\lambda\, X(x). \tag{12.6.16}$$

The initial and boundary conditions given in Equations (12.6.5), (12.6.6), and (12.6.13), give

$$T(0) > 0, \quad X(0) = 0, \quad X(L) = 0. \tag{12.6.17}$$

The solution to Equation (12.6.15) is

$$T(\lambda, t) = A(\lambda)e^{-\lambda Dt}, \tag{12.6.18}$$

where it has been indicated that the integration constant may be a function of λ, i.e.,

$$T(\lambda, 0) = A(\lambda). \tag{12.6.19}$$

Likewise, the solution to Equation (12.6.16) is

$$X(\lambda, x) = B_1(\lambda)\ \sin\left(\sqrt{\lambda}x\right) + B_2(\lambda)\ \cos\left(\sqrt{\lambda}x\right). \tag{12.6.20}$$

Now

$$X(\lambda, 0) = 0^{+} B_2(\lambda) \ \Rightarrow B_2(\lambda)\ = 0, \tag{12.6.21}$$

and

$$X(\lambda, L) = B_1(\lambda)\ \sin\left(\sqrt{\lambda}L\right), \tag{12.6.22}$$

gives

$$\sqrt{\lambda}L = n\pi \ \Rightarrow \lambda_n = \frac{\pi^2 n^2}{L^2},\ n = (1, 2, 3, \ldots) \tag{12.6.23}$$

Therefore, $X(\lambda, x)$ is

$$X_n(\lambda_n, x) = B_1(\lambda_n)\ \sin\left(\sqrt{\lambda_n}x\right), \tag{12.6.24}$$

and,

$$T_n(t) = T(\lambda_n, t) = A(\lambda_n)\, e^{-\lambda_n D t}, \tag{12.6.25}$$

where putting, these together gives the separation of variable solution

$$\begin{aligned} u_n(x, t) &= X_n(x)\ \ T_n(t) \\ &= b_n \sin\left(\frac{\pi n x}{L}\right)\ \exp\left[-\left(\frac{\pi^2 n^2 D}{L^2}\right)t\right]. \end{aligned} \tag{12.6.26}$$

In this expression

$$b_n = A(\lambda_n)\ B_1(\lambda_n). \tag{12.6.27}$$

Since the diffusion partial differential equation is linear, then sum of the individual solutions are solution. This means that a general solution takes the form

$$\begin{aligned} u(x, t) &= \sum_{n=1}^{\infty} u_n(x, t) \\ &= \sum_{n=1}^{\infty} b_n\ \sin\left(\frac{\pi n x}{L}\right)\exp\left[-\left(\frac{\pi^2 n^2}{L^2}\right)Dt\right]. \end{aligned} \tag{12.6.28}$$

Note that for $t = 0$, we have

$$\begin{aligned} u(x, 0) &= \sum_{n=1}^{\infty} b_n \sin\left(\frac{\pi n x}{L}\right) \\ &= f(x),\ \ 0 \le x \le L. \end{aligned} \tag{12.6.29}$$

Therefore, the above expansion is the sine Fourier series of $f(x)$, the initial condition for $u(x, t)$. The coefficients $\{b_n : n = 1, 2, 3, ...\}$ can be easily calculated and are given by the expression

$$b_n = \left(\frac{2}{L}\right)\int_0^L f(x) \sin\left(\frac{\pi n x}{L}\right) dx. \tag{12.6.30}$$

12.7 COMMENTS

This chapter has presented several models for the derivation of the heat/diffusion equation

$$u_t(x,t) = Du_{xx}(x,t). \tag{12.7.1}$$

These models start with assumed discrete microscopic models and end with continuous macroscopic extension, namely, the partial differential equation given in Equation (12.7.1). There is a great deal of ambiguity as to exactly what has been done, because various limits, such as

$$\Delta x \to 0, \Delta t \to 0, \frac{(\Delta x)^2}{\Delta t} = \text{constant}, \tag{12.7.2}$$

are taken, and it is not clear what Δx and Δt correspond to physically. In general, multiple models are created because no single model can explain or reproduce all of the properties of a particular system. Here, all the models give rise to the same macroscopic equation, i.e., the standard heat/diffusion. In a sense, this is somewhat surprising since the micro-level models are very crude representations of phenomena at these length and time scales.

A comparison of the micro-scale model

$$u_m^{k+1} = \left(\frac{1}{2}\right)u_{m+1}^k + \left(\frac{1}{2}\right)u_{m-1}^k, \tag{12.7.3}$$

with the heat/diffusion equation does not immediately indicate a direct connection between them. At this level of analysis, Equation (12.7.3) could be considered a finite difference discretization of Equation (12.7.3) with

$$R \equiv \frac{D\Delta t}{(\Delta x)^2} = \frac{1}{2}. \tag{12.7.4}$$

However, it is certainly clear that

$$u_m^k \neq u(x_m, t_k), \tag{12.7.5}$$

where $u(x,t)$ and u_m^k are, respectively, solutions to Equations (12.7.1) and (12.7.3). Thus, assuming the validity of Equation (12.7.3), its macroscopic extension, Equation (12.7.1), is only an approximation to it. Figure 12.4 provides a summary of this discussion.

In conclusion, both Equations (12.7.1) and (12.7.3) are phenomenological models. Both provide descriptions of heat/diffusion phenomena and are related, but not equivalent to each other. Both are consistent with the fundamental (classical) theories of physical theory, but are not directly derivable from them.

PROBLEMS

Section 12.5

1) Use the method of separation of variables to calculate a general solution to the realistic heat equation

$$\tau u_{tt} + u_t = Du_{xx}.$$

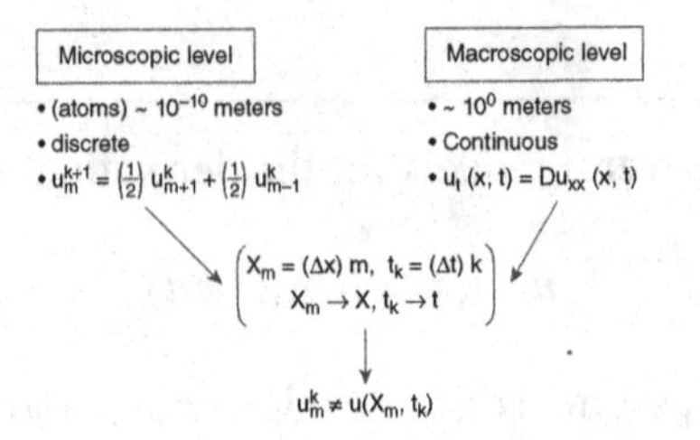

Figure 12.4: Interconnections of the microscopic and macroscopic levels.

2) For the standard heat equation

$$u_t = Du_{xx},$$

we can obtain a solution under the conditions

- $u(x, 0) = f(x)$ given, $0 \leq x \leq L$.
- $u(0, t) = u_0, u(L, t)\, u_L$, with $u(0, t) = f(0), u(L, t) = f(L)$.

However, for the relativistic heat equation, we also need $u_t(x, 0) = g(x)$ for $0 \leq x \leq L$. But, $g(x)$ is generally unknown. What can be done (if anything) to obtain a "reasonable" approximation to $g(x)$?

3) Does $u(x, 0) \geq 0$, for the realistic heat equation, that $u(x, t) \geq 0$, for $t > 0$?

Section 12.6

4) The heat/diffusion equation was derived from the relation

$$u_m^{k+1} = \left(\frac{1}{2}\right) u_{m+1}^k + \left(\frac{1}{2}\right) u_{m-1}^k.$$

Obtain a general solution to this equation using the separation of variables ansatz

$$u_m^k = C(k)\, F(m).$$

Compare this solution to the separation of variables representation of

$$u_t = Du_{xx}.$$

Remember that

$$t \to t_k = (\Delta t)\, k, x \to x_m = (\Delta x) m,$$

with

$$(\Delta x)^2 / \Delta t = \text{constant}.$$

NOTES AND REFERENCES

Sections 12.2, 12.3, and 12.4: Discussions on the derivation of the heat/diffusion partial differential equation and related issues are presented in the following books

(1) J. Crank, **The Mathematics of Diffusion**. (Clarendon Press, Oxford, 1975). See Sections 1.1, 1.2, and 1.3.

(2) R. Ghez, **A Primer of Diffusion Problems**, (Wiley, New York, 1988). See Chapter 1.

(3) J.D. Murray, **Mathematical Biology** (Springer-Verlag, Berlin, 1989). See Section 9.1.

Section 12.5: Serveral articles discussing the relativisitic heat equation and some nonlinear generalizations are

(4) D.C. Kelly, Diffusion: A relativistic approasch, **American Journal of Physics**, Vol. 36 (1968), 585-591.

(5) J. B. Keller, Diffusion at finite speed and random walks, **Proceedings of the National Academy of Sciences of the United States of America**, Vol. 101 (2004), 1120-1122.

(6) Y. Brenier, Extended Monge-Kantorovich theory, in M. Franca (editor), **Optimal Transportation and Applications** (Lecture Notes in Mathematics, Springer, Berlin, 2003), Vol. 1813, 91-121.

(7) E. Miller and A. Stern, Maximum principles for the relativistic heat equation, arXiv:1507.05030v1 [math.AP] 17 Jul 2015.

Note that the relativistic heat equation is a hyperbolic partial differential equation. A reference which discusses briefly this issue is

(8) Wikipedia: “Relativistic heat conduction.”

Section 12.6: The above indicated references by Crank (1975), Ghez (1988), and Murray (1989), collectively introduce and solve a broad class of problems involving diffusion and heat conduction. Similarity methods for both linear and nonlinear heat/diffusion processes are discussed in

(9) J.D. Logan, **Nonlinear Partial Differential Equations** (Wiley, New York, 1994). See Sections 4.1, 4.2, and 4.3.

(10) L. Dresner, **Similarity Solutions of Nonlinear Partial Differential Equations** (Pitman, Boston, 1983).

Appendix

A. ALGEBRAIC RELATIONS

A.1 Factors and Expansions

$$(a \pm b)^2 = a^2 \pm 2ab + b^2$$
$$(a \pm b)^3 = a^3 \pm 3a^2b + 3ab^2 \pm b^3$$
$$(a + b + c)^2 = a^2 + b^2 + c^2 + 2(ab + ac + bc)$$
$$(a + b + c)^3 = a^3 + b^3 + c^3 + 3a^2(b + c) + 3b^2(a + c) + 3c^2(a + b) + 6abc$$
$$a^2 - b^2 = (a - b)(a + b)$$
$$a^2 + b^2 = (a + ib)(a - ib), \quad i = \sqrt{-1}$$
$$a^3 - b^3 = (a - b)\left(a^2 + ab + b^2\right)$$
$$a^3 + b^3 = (a + b)\left(a^2 - ab + b^2\right)$$

A.2 Quadratic Equations

The quadratic equation

$$ax^2 + bx + c = 0$$

has the two solutions

$$x_1 = \frac{-b + \sqrt{b^2 - 4ac}}{2a},$$

$$x_2 = \frac{-b - \sqrt{b^2 - 4ac}}{2a}.$$

A.3 Cubic Equations

The cubic equation

$$x^3 + px^2 + qx + r = 0$$

can be reduced to the expression

$$z^3 + az + b = 0$$

by means of the substitution

$$x = z - \frac{p}{3},$$

where the constants a and b are

$$a = \frac{3q - p^2}{3},$$

$$b = \frac{3p^3 - 9pq + 27r}{27}.$$

New define A and B as

$$A = \left[-\left(\frac{b}{2}\right) + \left(\frac{b^2}{4} + \frac{a^3}{27}\right)^{\frac{1}{2}}\right]^{\frac{1}{3}},$$

$$B = \left[-\left(\frac{b}{2}\right) - \left(\frac{b^2}{4} + \frac{a^3}{27}\right)^{\frac{1}{2}}\right]^{\frac{1}{3}}.$$

The three roots of $z^3 + az + b = 0$ are

$$z_1 = A + B,$$

$$z_2 = -\left(\frac{A+B}{2}\right) + \sqrt{-3}\left(\frac{A-B}{2}\right),$$

$$z_3 = -\left(\frac{A+B}{2}\right) - \sqrt{-3}\left(\frac{A-B}{2}\right).$$

Let

$$\Delta = \frac{b^2}{4} + \frac{a^3}{27},$$

then for

(i) $\Delta > 0$: there will be one real root and two complex conjugate roots.

(ii) $\Delta = 0$: there will be three real roots, for which at least two are equal.

(iii) $\Delta < 0$: all roots are real and unequal.

A.4 Expansions of Selected Functions

- $\dfrac{1}{1-x} = 1 + x + x^2 + \cdots = \sum_{k=0}^{\infty} x^k, \quad |x| < 1$

- $(1+x)^q = 1 + qx + \dfrac{q(q-1)}{2!}x^2 + \cdots + \dfrac{q(q-1)\ldots(q-k+1)}{k!}x^k + \cdots$

The above series generally converges for $|x| < 1$.

- $e^x = \sum_{k=0}^{\infty} \dfrac{x}{k!}$

- $\cos x = \sum_{k=0}^{\infty} (-1)^k \frac{x^{2k}}{(2k)!}$
- $\sin x = \sum_{k=0}^{\infty} (-1)^k \frac{x^{2k+1}}{(2k+1)!}$
- $\mathrm{Ln}\,(1+x) = \sum_{k=1}^{\infty} (-1)^{k+1} \frac{x^k}{k}$

B. TRIGONOMETRIC RELATIONS

B.1 Fundamental Properties

$$(\sin\theta)^2 + (\cos\theta)^2 = 1$$

$$-1 \le \sin\theta \le +1, \qquad -1 \le \cos\theta \le +1$$

$$\sin(-\theta) = -\sin\theta, \qquad \cos(-\theta) = \cos\theta$$

$$\sin(\theta + 2\pi) = \sin\theta, \qquad \cos(\theta + 2\pi) = \cos\theta$$

$$\begin{cases} \sin(0) = 0, & \cos(0) = 1 \\ \sin\left(\frac{\pi}{2}\right) = 1, & \cos\left(\frac{\pi}{2}\right) = 0 \\ \sin(\pi) = 0, & \cos(\pi) = -1 \\ \sin\left(\frac{3\pi}{2}\right) = -1, & \cos\left(\frac{3\pi}{2}\right) = 0 \\ \sin(2\pi) = 0, & \cos(2\pi) = 1 \end{cases}$$

$$e^{i\theta} = \cos\theta + i\sin\theta, \qquad i = \sqrt{-1}$$

$$\sin\theta = \frac{e^{i\theta} - e^{-i\theta}}{2i}$$

$$\cos\theta = \frac{e^{i\theta} + e^{-i\theta}}{2i}$$

B.2 Sums of Angles

$$\sin(\theta_1 \pm \theta_2) = \sin\theta_1 \cos\theta_2 \pm \cos\theta_1 \sin\theta_2$$
$$\cos(\theta_1 \pm \theta_2) = \cos\theta_1 \cos\theta_2 \mp \sin\theta_1 \sin\theta_2$$

B.3 Other Trigonometric Relations

$$\sin\theta_1 \pm \sin\theta_2 = 2\sin\left(\frac{\theta_1 \pm \theta_2}{2}\right)\cos\left(\frac{\theta_1 \mp \theta_2}{2}\right)$$

$$\cos\theta_1 + \cos\theta_2 = 2\cos\left(\frac{\theta_1 + \theta_2}{2}\right)\cos\left(\frac{\theta_1 - \theta_2}{2}\right)$$

$$\cos\theta_1 - \cos\theta_2 = -2\sin\left(\frac{\theta_1 + \theta_2}{2}\right)\sin\left(\frac{\theta_1 - \theta_2}{2}\right)$$

$$\sin\theta_1\cos\theta_2 = \left(\frac{1}{2}\right)[\sin(\theta_1 + \theta_2) + \sin(\theta_1 - \theta_2)]$$

$$\cos\theta_1\sin\theta_2 = \left(\frac{1}{2}\right)[\sin(\theta_1 + \theta_2) - \sin(\theta_1 - \theta_2)]$$

$$\cos\theta_1\cos\theta_2 = \left(\frac{1}{2}\right)[\cos(\theta_1 + \theta_2) + \cos(\theta_1 - \theta_2)]$$

$$\sin\theta_1\sin\theta_2 = \left(\frac{1}{2}\right)[\cos(\theta_1 + \theta_2) - \cos(\theta_1 - \theta_2)]$$

B.4 Derivatives and Integrals

$$\begin{cases} \dfrac{d}{d\theta}\cos\theta = -\sin\theta, & \dfrac{d}{d\theta}\sin\theta = \cos\theta \\ \dfrac{d^2}{d\theta^2}\cos\theta = -\cos\theta, & \dfrac{d^2}{d\theta^2}\sin\theta = -\sin\theta \end{cases}$$

$$\begin{cases} \int\cos\theta\, d\theta = \sin\theta + C_1 \\ \int\sin\theta\, d\theta = -\cos\theta + C_2 \end{cases}$$

B.5 Powers of Trigonometric Functions

$$(\sin\theta)^2 = \left(\frac{1}{2}\right)(1 - \cos 2\theta)$$

$$(\cos\theta)^2 = \left(\frac{1}{2}\right)(1 + \cos 2\theta)$$

$$(\sin\theta)^3 = \left(\frac{1}{4}\right)(3\sin\theta - \sin 3\theta)$$

$$(\cos\theta)^3 = \left(\frac{1}{4}\right)(3\cos\theta + \cos 3\theta)$$

$$(\sin\theta)^4 = \left(\frac{1}{8}\right)(3 - 4\cos 2\theta + \cos 4\theta)$$

$$(\cos\theta)^4 = \left(\frac{1}{8}\right)(3 + 4\cos 2\theta + \cos 4\theta)$$

$$(\sin\theta)^5 = \left(\frac{1}{16}\right)(10\sin\theta - 5\sin 3\theta + \sin 5\theta)$$

$$(\cos\theta)^5 = \left(\frac{1}{16}\right)(10\cos\theta + 5\cos 3\theta + \cos 5\theta)$$

C. HYPERBOLIC FUNCTIONS

C.1 Definitions and Basic Properties

$$\text{hyperbolic sine: } \sinh(x) \equiv \frac{e^x - e^{-x}}{2}$$

$$\text{hyperbolic cosine: } \cosh(x) \equiv \frac{e^x + e^{-x}}{2}$$

It follows that

$$e^x = -\cosh(x) + \sinh(x),$$

$$e^{-x} = -\cosh(x) - \sinh(x),$$

and

$$[\cosh(x)]^2 - [\sinh(x)]^2 = 1.$$

C.2 Basic Properties

$$\cosh(-x) = \cosh(x), \quad \sinh(-x) = -\sinh(x)$$

$$\cosh(0) = 1, \quad \sinh(0) = 0$$

$$\operatorname*{Lim}_{x\to\pm\infty} \cosh(x) = +\infty$$

$$\operatorname*{Lim}_{x\to\pm\infty} \sinh(x) = \pm\infty$$

C.3 Derivatives and Integrals

$$\begin{cases} \dfrac{d}{dx}\cosh(x) = \sinh(x), & \dfrac{d}{dx}\sinh(x) = \cosh(x) \\ \dfrac{d^2}{dx^2}\cosh(x) = \cosh(x), & \dfrac{d^2}{dx^2}\sinh(x) = \sinh(x) \end{cases}$$

$$\begin{cases} \int \sinh(x)\,dx = \cosh(x) + C_1 \\ \int \cosh(x)dx = \sinh(x) + C_2 \end{cases}$$

C.4 Other Relations

$$\sinh(x \pm y) = \sinh(x)\cosh(y) \pm \cosh(x)\sinh(y)$$
$$\cosh(x \pm y) = \cosh(x)\cosh(y) \pm \sinh(x)\sinh(y)$$
$$\sinh(x) + \sinh(y) = 2\sinh\left(\frac{x+y}{2}\right)\cosh\left(\frac{x-y}{2}\right)$$
$$\cosh(x) + \cosh(y) = 2\cosh\left(\frac{x+y}{2}\right)\cosh\left(\frac{x-y}{2}\right)$$
$$\sinh(x) - \sinh(y) = 2\cosh\left(\frac{x+y}{2}\right)\sinh\left(\frac{x-y}{2}\right)$$
$$\cosh(x) - \cosh(y) = 2\sinh\left(\frac{x+y}{2}\right)\sinh\left(\frac{x-y}{2}\right)$$

C.5 Relations between Hyperbolic and Trigonometric Functions

$$\sinh(x) = -i\,\sin(ix),\; i = \sqrt{-1}$$
$$\cosh(x) = \cos(ix)$$

D. RELATIONS FROM CALCULUS

D.1 Differentiation

In the following, C_1 and C_2 are arbitrary constants. Further, the following notation is used

$$f'(x) = \frac{d}{dx} f(x).$$

$$\frac{d}{dx}[C_1 f(x) + C_2 g(x)] = C_1 f'(x) + C_2 g'(x)$$
$$\frac{d}{dx}[f(x)\,g(x)] = g(x) f'(x) + f(x)\,g'(x)$$
$$\frac{d}{dx} e^{f(x)} = f'(x)\,e^{f(x)}$$
$$\frac{d}{dx}\left[\frac{f(x)}{g(x)}\right] = \frac{g(x) f'(x) - f(x)\,g'(x)}{[g(x)]^2}$$
$$\frac{d}{dx}[f(g(x))] = f'(g(x))\,g'(x)$$

D.2 Integration by Parts

$$\int f(x)\, dg(x) = f(x)\, g(x) - \int g(x)\, df(x)$$

D.3 Differentiation of a Definite Integral with Respect to a Parameter

Let $f(x, t)$ and its derivative $\partial f(x,t)/\partial t$ be continuous in some domain in the $x - t$ plane which includes the rectangle

$$\psi(t) \leq x \leq \phi(t), \quad t_1 \leq t \leq t_2.$$

Assume $\psi(t)$ and $\phi(t)$ are defined and have continuous for derivatives for $t_1 \leq t \leq t_2$. Then for $t_1 \leq t \leq t_2$, it follows that

$$\frac{d}{dt}\int_{\psi(t)}^{\phi(t)} f(x, t)\, dx = f[\phi(t), t]\, \frac{d\phi}{dt} - f[\psi(t), t]\, \frac{d\psi}{dt} + \int_{\psi(t)}^{\phi(t)} \frac{\partial}{\partial t} f(x, t)\, dx.$$

D.4 Some Important Integrals

$$\int a dx = ax + c$$

$$\int x^p dx = \frac{x^{p+1}}{p+1} + c, \quad p \neq -1$$

$$\int \left(\frac{1}{x}\right) dx = \mathrm{Ln}\,|x| + c$$

$$\int e^x\, dx = e^x + c$$

$$\int \mathrm{Ln}(x)\, dx = x\mathrm{Ln}(x) - x$$

$$\int_0^\infty \frac{x^m\, dx}{x^n + a^2} = \frac{\pi\, a^{m-n+1}}{n \sin\left[\left(\frac{m+1}{n}\right)\pi\right]}, \quad 0 < m + 1 < n$$

$$\int_0^\infty e^{-ax} \cos(bx)\, dx = \frac{a}{a^2 + b^2}$$

$$\int_0^\infty e^{-ax} \sin(bx)\, dx = \frac{b}{a^2 + b^2}$$

$$\int_0^{\infty} e^{-ax^2} dx = \left(\frac{1}{2}\right) \sqrt{\frac{\pi}{a}}, \ a > 0$$

$$\int_0^{1} x^m [\mathrm{Ln}\,(x)]^n dx = \frac{(-1)^n n!}{(m+1)^{n+1}},$$

where $m > -1; \ n = 0, 1, 2, \ldots.$

$$\int_0^{2\pi} e^{i(m-n)\,x} \ dx = 2\pi \ \delta_{mn}$$

where δ_{mn} is the Kronecker "discrete" delta function. It is defined as follows

$$(m, n) = \text{integers}$$

$$\delta_{mn} = \begin{cases} 1, & m = n, \\ 0, & m \neq 0. \end{cases}$$

E. FOURIER SERIES

Consider a periodic function $f(x)$, of period $2L$,

$$f\,(x + 2L) = f\,(x)\,,$$

defined on $-\infty < x < \infty$. Assume that the following integrals exist

$$\int_0^{2L} f\,(x) \cos\left(\frac{k\pi x}{L}\right) dx, \quad \int_0^{2L} f\,(x) \sin\left(\frac{k\pi x}{L}\right) dx,$$

for $k = 0, 1, 2, \ldots$, and the formal Fourier series of $f\,(x)$ on $0 < x < 2L$ is given by the expression

$$f\,(x) \sim \frac{a_0}{2} + \sum_{k=1}^{\infty} \left[a_k \cos\left(\frac{k\pi x}{L}\right) + b_k \sin\left(\frac{k\pi x}{L}\right) \right],$$

where

$$a_k = \left(\frac{1}{L}\right) \int_0^{2L} f\,(x) \cos\left(\frac{k\pi x}{L}\right) dx,$$

$$b_k = \left(\frac{1}{L}\right) \int_0^{2L} f\,(x) \ \sin\left(\frac{k\pi x}{L}\right) dx.$$

Definition .1 *A function $f(x)$ is said to be* **piecewise continuous** *on a finite interval, $a \leq x \leq b$, if this interval can be partitioned into a finite number of subintervals such that $f(x)$ is continuous in the interior of each of the subintervals and $f(x)$ has finite limits as x approaches either end point of each subinterval from its interior.*

Definition .2 *A function $f(x)$ is said to be* **piecewise smooth** *on a finite interval, $a \leq x \leq b$, if both $f(x)$ and $f'(x)$ are piecewise continuous on $a \leq x \leq b$.*

Theorem .1 *Let $f(x)$ be piecewise smooth on the interval, $0 < x < 2L$. Then its Fourier series is*

$$f(x) = \frac{a_0}{2} + \sum_{k=1}^{\infty} \left[a_k \cos\left(\frac{k\pi x}{L}\right) + b_k \sin\left(\frac{k\pi x}{L}\right) \right].$$

This series converges at every point x, in the interval $0 < x < 2L$, to the value

$$\frac{f(x^+) + f(x^-)}{2},$$

where $f(x^+)$ and $f(x^-)$ are, respectively, the right- and left-hand limits of f at x.
If f is continuous at x, then the Fourier series of f at x converges to $f(x)$.

F. EVEN AND ODD FUNCTIONS

Let the functions $f(x)$ and $g(x)$ be defined on the symmetric interval, where a may be unbounded.

(1) A function $f(x)$ is an even function on the interval $(-a, a)$, if and only if

$$f(-x) = f(x).$$

(2) A function $f(x)$ is an odd function on the interval $(-a, a)$, if and only if

$$f(-x) = -f(x).$$

(3) Given an arbitrary function $g(x)$, defined on the interval $(-a, a)$, then it can be written as

$$g(x) = g^{(+)}(x) + g^{(-)}(x),$$

where

$$g^{(+)}(x) = \frac{g(x) + g(-x)}{2},$$
$$g^{(-)}(x) = \frac{g(x) - g(-x)}{2}.$$

The functions $g^{(+)}(x)$ and $g^{(-)}(x)$ are, respectively, the even and odd parts of $g(x)$.

(4) Let $f(x)$ and $g(x)$ be both even functions. Then, $h(x) = f(x)\ g(x)$ is an even function.

(5) If $f(x)$ and $g(x)$ are both odd functions, then $h(x) = f(x)\ g(x)$ is an even function.

(6) If $f(x)$ is an even function and $g(x)$ is an odd function, then $h(x) = f(x)\ g(x)$ is an odd function.

(7) Let $f(x)$ be an even function, integrable on the interval $(-a, a)$, then

$$\int_{-a}^{a} f(x)\,dx = 2\int_{0}^{a} f(x)\,dx.$$

(8) If $f(x)$ is an odd function, integrable on the interval $(-a, a)$, then

$$\int_{-a}^{a} f(x)\,dx = 0.$$

(9) Let $f(x)$ be an even (odd) function over the interval $(-a, a)$, then if the derivative exists, it is odd (even) on this same interval, i.e.,

$$f(x) \text{ even} \Rightarrow \frac{df(x)}{dx} \text{ odd},$$
$$f(x) \text{ odd} \Rightarrow \frac{df(x)}{dx} \text{ even}.$$

G. SOME NONSTANDARD BUT IMPORTANT FUNCTIONS

G.1 Absolute Value Function

Let a be a nonzero real number. The absolute value of a is defined to be

$$|a| = \begin{cases} a, & \text{if } a > 0, \\ 0, & \text{if } a = 0, \\ -a, & \text{if } a < 0. \end{cases}$$

An alternative, but sometimes, useful definition is

$$|a| = \sqrt{a^2}.$$

The absolute value function has the following properties:

(i) $|a| \geq 0$

(ii) $|-a| = |a|$

(iii) $|ab| = |a|\ |b|$

(iv) $\left|\frac{a}{b}\right| = \frac{|a|}{|b|}$

(v) $|a^n| = |a|^n$

(vi) $|a + b| \leq |a| + |b|$

The above properties generalized to the situation where a and b are functions. The graph of $|x|$ vs x is given in Figure G.1.

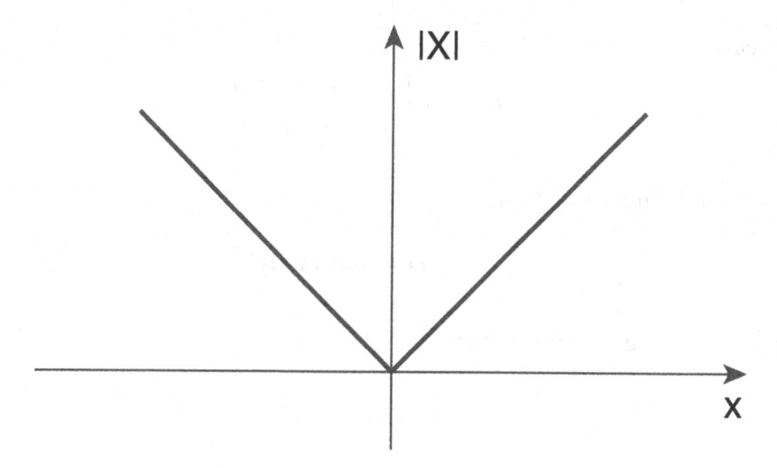

Figure G.1: Plot of the absolute value function, $|x|$.

G.2 Sign and Theta Functions

The sign and theta functions are defined respectively, as follows

$$\text{sign}(x) \equiv \begin{cases} 1, & \text{if } x > 0, \\ -1, & \text{if } x < 0, \end{cases}$$

$$\theta(x) \equiv \begin{cases} 1, & \text{if } x > 0, \\ 0, & \text{if } x < 0. \end{cases}$$

Both functions are discontinuous at $x = 0$. However, depending on the circumstances, the following values are generally choosen

$$\text{sign}(0) = 0, \quad \theta(0) = \frac{1}{2}.$$

Often the theta function is called the **Heaviside step function** and denoted $H(x)$, i.e.,

$$H(x) = \theta(x).$$

Figure G.2 gives plots of sign (x) and $\theta(x)$ vs x.
Note that the following relations hold

$$\text{Sign}(x) = \theta(x) - \theta(-x),$$
$$\text{Sign}(x) = \frac{d}{dx}|x|, \ x \neq 0.$$

G.3 Ramp Function

The ramp function, $R(x)$, may be defined in several different, but equivalent ways:

(i) As a piecewise function

$$R(x) = \begin{cases} x, & \text{if } x \geq 0, \\ 0, & \text{if } x < 0. \end{cases}$$

(ii) From use of the max-function

$$R(x) = \max(x, 0).$$

(iii) Using the absolute value function

$$R(x) = \frac{x + |x|}{2}.$$

(iv) Using the theta function

$$R(x) = x\theta(x).$$

(v) As an integral of the theta function

$$R(x) = \int_{-\infty}^{x} \theta(z)\, dz.$$

Figure G.3 gives a plot of the Ramp function.

G.4 Boxcar or Rectangular Function

The boxcar function, $B(x)$, is zero over the entire real line, except for an interval, (a, b), where it takes the constant value one. A plot of this function is given in Figure G.4.

The boxcar function may be defined as follows

$$B(x) = B(x; a, b) \equiv \begin{cases} 1, & \text{if } a < x < b, \\ 0, & \text{all other values of } x. \end{cases}$$

It may also be expressed in terms of the theta function

$$B(x) = \theta(x - a) - \theta(x - b).$$

G.5 Triangular Function

The triangular function also goes under several other names: triangle function, hat function, and tent function. Its depiction is given in Figure G.5.

Expressed as a piecewise function, the triangular function takes the form

$$Tri(x) = \begin{cases} 1 - |x|, & |x| < 1, \\ 0, & \text{for all other } x \text{ values}. \end{cases}$$

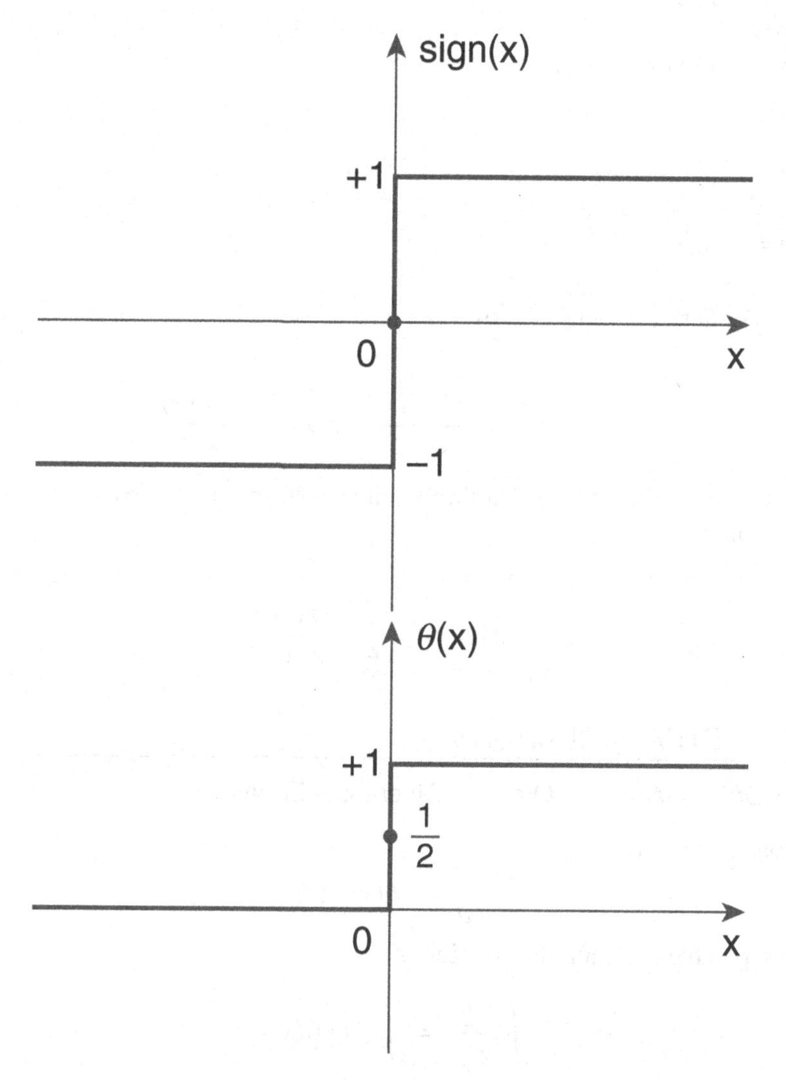

Figure G.2: (a) Plot of sign (x) vs x. (b) Plot of $\theta(x)$ vs x.

G.6 Delta Function

A function, $\delta(x)$, is called a "delta function" if it has the three properties:

(i) $\int\limits_{-\infty}^{\infty} \delta(x)\, dx = 1,$

(ii) $\delta(x) = 0, \quad \text{if } x = 0,$

(iii) $\int\limits_{-\infty}^{\infty} \delta(x)\, f(x)\, dx = f(0).$

This function, $\delta(x)$, also has the following properties:

(a) The delta function is even, i.e.,

$$\delta(-x) = \delta(x),$$

(b) $\int\limits_{-\infty}^{\infty} \delta(x-a) f(x)\, dx = f(a)$,

(c) $\int\limits_{-\infty}^{\infty} \delta(ax) f(x)\, dx = \frac{1}{|a|} f(0)$,

(d) $x\delta'(x) = -\delta(x)$

(e) $\int\limits_{-\infty}^{\infty} \delta^{(n)}(x) f(x)\, dx = (-1)^n f^{(n)}(0)$,

where

$$\delta^{(n)}(x) \equiv \frac{d^n \delta(x)}{dx^n}, \quad f^n(x) \equiv \frac{d^n f(x)}{dx^n}.$$

(f) Let $g(x)$ be a function having a finite number, N, of simple zeros at $(x_1, x_2, \ldots, x_n)$. We then have

$$\delta[g(x)] = \sum_{k=1}^{N} \frac{\delta(x-x_k)}{|g'(x_k)|}.$$

H. DIFFERENTIAL EQUATIONS

H.1 First-Order, Separable Ordinary Differential Equations

An equation of the form

$$\frac{dy}{dx} = f(x)\, g(y)$$

is called a separable equation. Its solution is

$$\int \frac{dy}{g(y)} = \int f(x)\, dx = c$$

where C is an arbitrary constant.

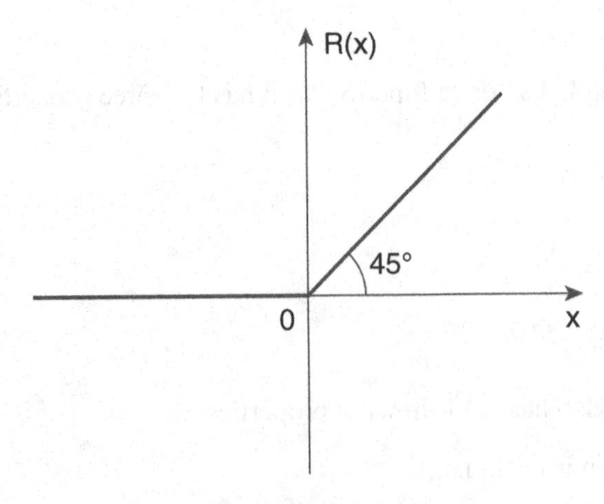

Figure G.3: Plot of $R(x)$ vs x.

H.2 General Linear, First-Order, Ordinary Differential Equation

This type of equation takes the form

$$\frac{dy}{dx} + P(x)\,y = Q(x)$$

where $P(x)$ and $Q(x)$ are given. The general solution is

$$y(x) = Ce^{-\int P(x)dx} + e^{-\int P(x)dx} \int e^{\int P(x)dx} Q(x)\,dx$$

where c is an arbitrary constant.

H.3 Bernoulli Equations

A commonly occurring first-order, nonlinear differential equation is the Bernoulli equation

$$\frac{dy}{dx} + P(x)\,y = Q(x)\,y^n, \quad n \neq 1,$$

where $P(x)$, $Q(x)$, and N are specified. The nonlinear transformation

$$u(x) = y^{(1-n)}$$

reduces the Bernoulli equation to the form

$$\frac{du}{dx} + P_1(x)\,u = Q(x),$$

which is linear in $u(x)$ and can solved by the method presented in section H.2.

H.4 Linear, Second-Order, Homogeneous Differential Equations with Constant Coefficients

These type of equations appear often in many applications and take the form

$$a_0 \frac{d^2y}{dx^2} + a_1 \frac{dy}{dx} + a_2 y = 0,$$

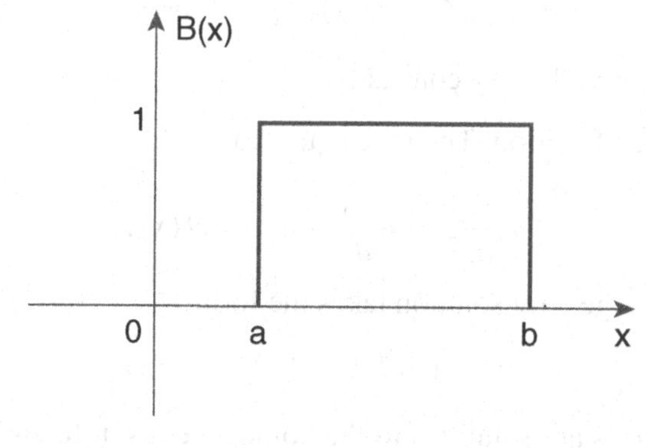

Figure G.4: Plot of the boxcar function.

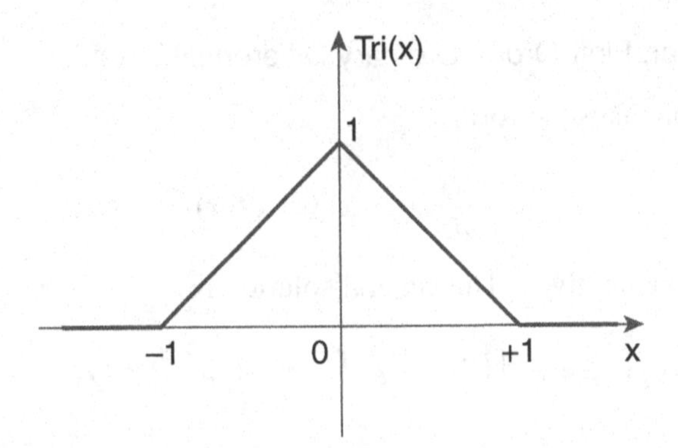

Figure G.5: Plot of the triangular function, Tri (x) vs x.

where the coefficients $(a_0,\ a_1,\ a_2)$ are real constants. The equation

$$a_0 m^2 + a_1 m + a_2 = 0,$$

is called the characteristic equations, and its two roots, m_1 and m_2, determine the solutions of the differential equations as follows:

(i) Let m_1 and m_2 be real and distinct, i.e., $m_1 \neq m_2$. The general solution is

$$y(x) = c_1 e^{m_1 x} + c_2 e^{m_2 x},$$

where C_1 and C_2 are arbitrary constants.

(ii) Let m_1 and m_2 be complex conjugates of each other, i.e.,

$$m_1 = m_2^* = a + bi, \quad i = \sqrt{-1}.$$

The solution is

$$yx = \begin{cases} Ae^{ax} \quad \text{or} \quad \cos(bx + B), \\ e^{ax}\left[C_1 \cos(bx) + C_2 \sin(bx)\right], \end{cases}$$

where $(A,\ B,\ C_1, C_2)$ are real.

(iii) If $m_1 = m_2$, then the general solution is

$$y(x) = (C_1 + C_2 x)e^{mt}, \qquad m = m_1 = m_2$$

where $(C_1,\ C_2)$ are arbitrary constants.

Let $Q(x)$ be a given function. Then the equation

$$a_0 \frac{d^2 y}{dx^2} + a_1 \frac{dy}{dx} + a_2 y = Q(x),$$

is inhomogeneous. The general solution takes the form

$$y(x) = C_1 e^{m_1 x} + C_2 e^{m_2 x} + v(x),$$

where the first-two terms are solutions to the homogeneous differential equation and $v(x)$ is a **particular solution** to the inhomogeneous equation.

The following two rules allow the determination of the particular solution:

(1) **Rule 1:** Let no term in $Q(x)$ be the same as in

$$u(x) = C_1 e^{m_1 x} + C_2 e^{m_2 x}.$$

For this case, $v(x)$ will be linear combination of the terms in $Q(x)$ and all their independent derivatives.

(2) **Rule 2:** Let $Q(x)$ contain a term that, ignoring constant coefficients, is x^k times a term $u_1(x)$ appearing in $u(x)$, where $k = 0, 1, 2, \ldots$. For this case, a particular solution, $v(x)$, will be a linear combination of $x^{k+1} u_1(x)$ and all its linearly independent derivatives that are not contained in $u(x)$.

Note that if $Q(x)$ contains terms that do not give rise to a finite set of linearly independent derivatives, then no general method exists to directly calculate the particular solution.

I. LINEARIZATION OF CERTAIN TYPES OF NONLINEAR DIFFERENTIAL EQUATIONS

There are several types or classes of differential equations that can be exactly linearized. For example, see Appendix H.3, which discusses how this is done for Bernoulli equations

$$\frac{dy}{dx} + P(x)y = Q(x)y^n, \quad n \neq 1.$$

I.1 Riccati Equation

The nonlinear Riccati differential equation is

$$y' = q_0(x) + q_1(x)y + q_3(x)y^2,$$

where the "prime" notation is used for the derivative

$$y' = \frac{dy(x)}{dx}.$$

If $q_2(x)$ is non-zero and has a first derivative, then the nonlinear transformation

$$v(x) = q_2(x)y(x),$$

gives

$$v' = v^2 + R(x)v + S(x),$$

where

$$S(x) = q_0(x)q_2(x), \qquad R(x) = q(x) + \frac{q_2'(x)}{q_2(x)}.$$

Now, define $u(x)$ as

$$\begin{aligned} v(x) &= -\frac{u(x)'}{u(x)} \\ &= -[\text{Ln } u(x)]'. \end{aligned}$$

It follow that $u(x)$ is a solution to the following second-order, linear differential equation

$$u'' - R(x)u' + S(x)u = 0,$$

and the solution to the original Riccati equation is

$$\begin{aligned} y(x) &= -\frac{u'(x)}{q_2(x)u(x)} \\ &= -\left[\frac{1}{q_2(x)}\right][\text{Ln}\, u(x)]'. \end{aligned}$$

I.2 Quadratic Nonlinearities: Two Special Cases

Consider the following nonlinear, ordinary differential equation

$$\frac{dy}{dx} = a(x)y + b(x)y^2,$$

where $a(x)$ and $b(x)$ are specified. The change of variables

$$y = u^{-1},$$

gives the linear differential equation

$$\frac{du}{dx} = -a(x)u - b(x).$$

Now consider the nonlinear, first-order partial differential equation, $u = u(x, t)$,

$$u_t + a(x,t)u_x = b(x,t)u + c(x,t)u^2, \tag{I.1}$$

$$u_t \equiv \frac{\partial u(x,t)}{\partial t}, \quad u_x \equiv \frac{\partial u(x,t)}{\partial x},$$

and $(a(x,t),\ b(x,t),\ c(x,t))$ are specified. The change of variables

$$u(x,t) = v(x,t)^{-1},$$

gives the following linear first-order, partial differential equation

$$v_t + a(x,t)v_x = -b(x,t)v - c(x,t).$$

I.3 Square-Root Nonlinearities

The first-order, nonlinear differential equation

$$\frac{dy}{dx} = a(x) + b(x)\sqrt{y},$$

can be linearized by means of the variable change

$$u(x) = \sqrt{y(x)},$$

and gives

$$\frac{du}{dt} = \left(\frac{1}{2}\right) a(x) u + \frac{1}{2} b(x),$$

which is a first-order, linear, inhomogeneous differential equation.

For the first-order, nonlinear partial differential equation

$$u_t + a(x,t) u_x = b(x,t) u + c(x,t) \sqrt{u},$$

the transformation

$$v(x,t) = \sqrt{u(x,t)},$$

gives

$$v_t + a(x,t) v_x = \left(\frac{1}{2}\right) b(x,t) v + \left(\frac{1}{2}\right) c(x,t).$$

Bibliography

BOOKS

D. Basmadjian, **Mathematical Modeling of Physical Systems** (Oxford University Press, Oxford, 2003).

F. Brauer and C. Castillo-Chárez, **Mathematical Models in Population Biology and Epidemiology** (Springer, New York, 2001).

M. Braun, **Differential Equations and Their Applications, Fourth Edition** (Springer, New York, 1992).

T. P. Dreyer, **Modelling with Ordinary Differential Equations** (CRC Press, Boca Raton, 1993).

L. Edelstein-Keshet, **Mathematical Models in Biology** (McGraw-Hill, New York, 1987).

G. Fulford, P. Forrester, and A. Jones, **Modelling with Differential and Difference Equations** (Cambridge University Press, Cambridge, 1997).

F. R. Giordane and M. Weir, **Differential Equations: A Modeling Approach** (Addison-Wesley Longman, Upper Saddle River, NJ, 1991).

R. Haberman, **Mathematical Models** (Society for Industrial and Applied Mathematics, Philadelphia, 1998).

D. Hughes-Hallett, et al., **Applied Calculus, 5th Edition** (Wiley, Hoboken, NJ, 2013).

N.H. Ibragimov, **Practical Course in Differential Equations and Mathematical Modelling** (World Scientific Publishing, Singapore, 2010).

J. Mazumdar, **An Introduction to Mathematical Physiology and Biology** (Cambridge University Press, Cambridge, 1989).

M. Mesterton-Gibbons, **A Concrete Approach to Mathematical Modelling** (Wiley-Interscience, New York, 1995).

R.E. Mickens, **Difference Equations: Theory, Applications and Advanced Topics, Third Edition** (CRC Press, Boca Raton, 2015).

R.E. Mickens, **Mathematical Methods for the Natural and Engineering Sciences, Second, Edition** (World Scientific Publishing, Singapore, 2017).

L.A. Segel, **Modeling Dynamic Phenomena in Molecular and Cellular Biology** (Cambridge University Press, Combridge, 1984).

D.R. Shier and K.T. Wallenius, **Applied Mathmatical Modeling** (Chapman and Hall/CRC, Boca Raton, 2000).

T. Witelski and M. Bowen, **Methods of Mathematical Modelling** (Springer, New York, 2015).

HANDBOOKS

H.J. Bartsch, **Handbook of Mathematical Formulas** (Elsevier Science and Technology Books, Boston, 1974).

I.N. Bronschten, K.A. Semendyayer, G. Musiol, and H. Mühlig, **Handbook of Mathematics, Sixth Edition** (Springer, Heidelberg, 2015).

CRC Standard Mathematical Tables and Formulas, 33rd Edition, D. Zwillinger, editor (CRC Press, New York, 2018).

I.S. Gradshteyen and I.M. Ryzhik, **Table of Integrals, Series, and Products**, D. Zwillinger, editor (Elsevier, Boston, 2015).

A. Jeffrey and H. Dai, **Handbook of Mathematical Formulas and Integrals**, 4th Edition (Academic Press, New York, 2008).

G.A. Korn and T.M. Korn, **Mathematical Handbook for Scientists and Engineers** (Dover reprint edition, 2000; McGraw-Hill, New York, 1968).

NIST Handbook of Mathematical Functions, edited by F.J. Olver, D.W. Lozier, R.F. Boisvert, and C.W. Clark (Cambridge University Press, Cambridge, 2010).

A.D. Polyanin and A.V. Manzhirov, **Handbook of Mathematics for Engineers and Scientists** (Chapman and Hall/CRC Press, Boca Raton-London, 2006).

A.D. Polyanin and V.F. Zaitsev, **Handbook of Exact Solutions for Ordinary Differential Equations, 2nd Edition** (Chapman and hall/CRC Press, Boca Raton, 2003).

M.R. Spiegel, **Schaum's Mathematical Handbook of Formulas and Tables** (McGraw-Hill, New York, 1998).

D. Zwillinger, **Handbook of Differential Equations, 3rd Edition** (Academic Press, Boston, 1997).

WEBSITES

The following website provides extensive information on both ordinary and partial differential equations, integral and functional equations, and many other types of mathematical equations and functions:

EqWorld, The World of Mathematical Equations.

Another extensive website is

`http://dlmf.nist.gov`

which gives the

NIST Digital Library of Mathematical Functions (DLMF).

Index

Printed in the United States
by Baker & Taylor Publisher Services